AN APPROACH TO MEDIUM-TERM COASTAL
MORPHOLOGICAL MODELLING

An Approach to Medium-term Coastal Morphological Modelling

DISSERTATION
Submitted in fulfilment of the requirements of
the Board for Doctorates of Delft University of Technology
and of the Academic Board of the UNESCO-IHE Institute for Water Education
for the Degree of DOCTOR
to be defended in public
on Thursday, 4 June 2009 at 14:00 hours
in Delft, The Netherlands

by

Giles Ransom LESSER
born in Wellington, New Zealand

Bachelor of Engineering, University of Canterbury, New Zealand
Master of Science, IHE Delft.

This dissertation has been approved by the supervisor:
  Prof. dr. ir. J.A. Roelvink

Members of the Awarding Committee:
  Chairman   Rector Magnificus, TU Delft, the Netherlands
  Prof. dr. R.A.M. Meganck   Vice-Chairman, UNESCO-IHE, the Netherlands
  Prof. dr. ir. J.A. Roelvink UNESCO-IHE / TU Delft, Supervisor
  Prof. dr. ir. H.J. de Vriend TU Delft / Univ. Twente, the Netherlands
  Prof. dr. ir. G.S. Stelling TU Delft, the Netherlands
  Prof. dr. ir. M.J.F. Stive   TU Delft, the Netherlands
  Prof. dr. ir. L.C. van Rijn University of Utrecht, the Netherlands
  Dr. G. Gelfenbaum   US Geological Survey, USA

CRC Press/Balkema is an imprint of the Taylor & Francis Group, an informa business

Published by:
CRC Press/Balkema
PO Box 447, 2300 AK Leiden, The Netherlands
e-mail: Pub.NL@taylorandfrancis.com
www.crcpress.com - www.taylorandfrancis.co.uk - www.ba.balkema.nl

Cover Image: Computed long-term (residual) sand transport patterns around Empire Spit, North Cove, Willapa Bay, WA, USA. Background photograph courtesy of Aero-Metric Inc.

ISBN 978-0-415-55668-2 (Taylor & Francis Group)

All models are wrong. Some models are useful.
- George Box

*If you put tomfoolery into a computer, nothing comes out of it but tomfoolery.
But this tomfoolery, having passed through a very expensive machine, is somehow
ennobled and no-one dares criticize it.*
- Pierre Gallois

*The most overlooked advantage to owning a computer is that if they foul up
there's no law against wacking them around a little.*
- Joe Martin, Porterfield

Abstract

Process-based numerical modelling of sediment transport and morphological change is widely recognised as a valuable approach for understanding and predicting coastal morphological developments. In practice, state-of-the-art computer models are one- or two-dimensional (depth-averaged) and have a limited ability to model many of the important three-dimensional phenomena found in nature. Furthermore, input reduction and morphological acceleration techniques must usually be applied in order to extend the useful horizon of process-based morphological models from hydrodynamic time scales of minutes to hours to the generally more relevant morphological time scales of years to decades.

This thesis presents the implementation of fully three-dimensional sediment transport and morphological updating formulations within the proven Delft3D three-dimensional (hydrostatic, free surface) flow solver. The thesis briefly discusses the operation of the Delft3D-FLOW module, presents the formulations used to model both suspended and bed-load transport of non-cohesive sediment, and describes the implementation of a morphological updating scheme which incorporates novel approaches to morphological acceleration and dry bank erosion. Approaches used to model the three-dimensional effects of waves on coastal hydrodynamics and of three-dimensional currents on waves are also discussed.

Following the details of the implementation, the results of several validation studies are presented. The model is shown to perform well in a series of simplified theoretical, laboratory, and full scale test cases. Application of the model and acceleration techniques to the complex and dynamic entrance to Willapa Bay, WA, USA is also discussed. Model processes are validated against the results of an extensive field measurement campaign, and diagnostic morphological model simulations are performed for two periods of contrasting morphological development. Application of the model is found to be very beneficial as part of a multifaceted analysis of the morphology of Willapa Bay, however the morphological model predictions themselves are found to only partially reproduce the observed morphological changes.

Input reduction and morphological acceleration techniques used to perform 5-year simulations of Willapa Bay are critically analysed and a commonly used method of tidal input reduction is found to have systematic errors when applied to coastal environments containing significant diurnal tidal energy. A new generally applicable method to select a representative morphological tide for such environments is presented and is shown to perform well compared to brute force simulations. Errors introduced by the input reduction and morphological acceleration techniques applied to Willapa Bay are quantified and found to be negligible compared to errors due to shortcomings of the sediment transport formulations contained in the model. Several areas are identified for further advancement of coastal morphological modelling tools and techniques.

Lesser, G.R., 2009. *An approach to medium-term coastal morphological modelling*. PhD thesis, UNESCO-IHE & Delft Technical University, Delft. CRC Press/Balkema. ISBN 978-0-415-55668-2.

Samenvatting

Proces gebaseerd numeriek modelleren van sedimenttransport en morfologische veranderingen is breed erkend als een bruikbare methode voor het begrijpen en voorspellen van kustnabije morfologische ontwikkelingen. In de praktijk zijn 'state-of-the-art' computermodellen één- of tweedimensionaal (diepte gemiddeld) en kunnen zij maar beperkt de belangrijke driedimensionale effecten, zoals die zich voordoen in de natuur, modelleren. Ook is het vaak noodzakelijk om de hoeveelheid invoergegevens te beperken en om morfologische processen kunstmatig te versnellen om de tijdshorizon van morfologische modellen uit te breiden van de hydrodynamische tijdschalen van minuten tot uren naar de in het algemeen meer relevante morfologische tijdschalen van jaren tot decennia.

Dit proefschrift richt zich op de implementatie van volledig driedimensionale bodemveranderingsformuleringen in de bestaande Delft3D driedimensionale (hydrostatisch, vrij vloeistofoppervlak) stromingsmodule. Het proefschrift geeft een korte beschrijving van de werking van de Delft3D-Flow module, presenteert vervolgens de formuleringen zoals gebruikt voor het modelleren van zowel gesuspendeerd als bodemtransport van niet-cohesief sediment en beschrijft de implementatie van een bodemveranderingsschema dat gebruik maakt van nieuwe methodes voor het modelleren van morfologische versnelling en (rivier)bank erosie. Ook beschreven zijn de gebruikte methodes voor het modelleren van de driedimensionale effecten van golven op de kustnabije waterbeweging en de effecten van driedimensionale stromingen op golven.

Na de details van de implementatie worden de resultaten van verscheidene validatiestudies gepresenteerd. Aangetoond is dat het model goed presteert in een reeks vereenvoudigde theoretische (laboratorium) en praktijk testgevallen. Ook beschreven zijn de toepassing van het model en de versnellingstechnieken op het complexe en dynamische zeegat van Willapa Bay, WA, USA. De modelprocessen zijn gevalideerd door middel van uitgebreide veldmetingen, en 'diagnostische' morfologische modelsimulaties zijn uitgevoerd voor twee contrasterende periodes van morfologische ontwikkeling. Het model is bijzonder geschikt wanneer toegepast als onderdeel van een uitgebreide analyse van de morfologie van Willapa Bay. Het is echter gebleken dat de morfologische modelvoorspellingen maar gedeeltelijk de waargenomen morfologische ontwikkelingen reproduceren.

De reductie van invoergegevens en de morfologische versnellingstechnieken, zoals gebruikt voor de 5-jarige simulatie van Willapa Bay, zijn kritisch geanalyseerd. Het is aangetoond dat gebruik van een veel voorkomende methode voor het reduceren van getij invoergegevens, resulteert in systematische verschillen als het wordt toegepast op kustgebieden met significante enkeldaagse getijcomponenten. Een nieuwe algemene methode voor zulke kustgebieden is gepresenteerd en het is aangetoond dat deze methode goed presteert vergeleken met 'brute force' (brute kracht) simulaties. Onnauwkeurigheden, zoals geïntroduceerd door de toepassing van invoerreductie en morfologische versnellingstechnieken, zijn gekwantificeerd en zijn klein vergeleken met de onnauwkeurigheden in de sedimenttransportfor-muleringen in het model. Verscheidene gebieden voor verdere verbetering van kustmorfologische modelleringsgereedschappen en -technieken zijn geïdentificeerd.

Lesser, G.R., 2009. *An approach to medium-term coastal morphological modelling*. PhD thesis, UNESCO-IHE & Delft Technical University, Delft. CRC Press/Balkema. ISBN 978-0-415-55668-2.

Contents

Acknowledgements

This thesis is the fruit of eight years of on-and-off labour which took place on three continents. I am very fortunate to have had the opportunity to live, work, and play so widely and to have met and learned from so many people along the way. I consider myself very fortunate to have been a rolling stone and to have gathered more than a little moss.

This PhD study would not have been possible without the financial support of Delft Hydraulics, UNESCO-IHE, and the Technical University of Delft who have helped to fund the required time, materials, and travel. The remarkable support provided by the USGS, who have continued to grant remote access to a computer cluster at Menlo Park to complete those "last few" simulations, is also very gratefully acknowledged. This thesis has been written alongside my day job. Peter O'Brien and all at OMC, thank you for being supportive of my PhD study and allowing me to take time away from work. My thanks also to Chris Hens from OMC for dusting off his own "rusty" Dutch in order to translate my abstract.

The development of the "online" sediment transport model continues the direction suggested by Dano Roelvink for my Master's thesis and it is due, to a large extent, to his foresight and experience that this direction proved so successful. Dano you have the uncanny knack of spotting the tiniest glimpse of silver lining in any cloud, but you don't hesitate to point out rubbish when you see it. To be critical yet motivational is an art to which I aspire. Thank you for keeping this study alive by being both.

To the rest of the Delft Hydraulics crew who freely and willingly shared their knowledge and were happy to debate everything from wave-current interaction to the idiosyncrasies of the Dutch and English languages, thank you for so openly welcoming a foreigner into your midst. The memories and experience I gained in the Netherlands will last me a lifetime.

Guy Gelfenbaum thank you for arranging for my stay as Visiting Scientist at USGS Menlo Park and instructing me in the ways of field data collection. It is clear that without having the opportunity to test the model against a real-life application, and the time to investigate all the questions that arose, this thesis would never have taken shape. Thank you for teaching me that the most useful and interesting research usually takes place when things don't work out the way you had planned.

Laura Landerman (née Kerr). What can I say? A lot of water has passed under both our bridges since we innocently (and ignorantly) debated the reproductive rites of the delicious crabs that would sometimes come aboard with instruments in Willapa Bay. Thanks for being bridesmaid at our wedding (and getting Nate to drive the car). Thanks for the description of the Willapa experiment that I have borrowed from in Chapter 4 and, most of all, thanks for having that Jersey accent. Just talking to you on the phone still cracks me up.

Lauren, thank you for carrying the load while my mind was elsewhere. To have, and to have to take care of, three children while your husband has the perfect excuse for disappearing at critical moments just simply isn't fair. Thank you for encouraging me to complete this PhD. Without your encouragement and support this thesis would never have been completed.

In many ways these acknowledgements have missed the people who have been the most essential. Eight years is far too long to live without making some pretty good friends along the way. To "Dutch" drinking and sailing friends, to "American" biking and baby-making friends, and to Australian... no, strike that. You made it a pleasure, thank you.

Melbourne, November 2008.

Chapter 1

Introduction

1.1 Background

Coastal (geo)morphology is the study of the changing shape of coastal landforms and is of critical importance to the development of human infrastructure on or near the coast. Poor understanding of coastal processes and potential coastal morphological change leads to poor decisions regarding coastal management and development. Present understanding of the coastal processes responsible for morphological change is considerable, based on decades of painstaking scientific research, but is still far from complete (e.g. Stive and Wang, 2003). The concept of separation of scales is also critical (de Vriend et al., 1993; Stive and Wang, 2003) as the prospect of describing the movement of every grain of sand on several kilometres of beach or in an entire estuary due to every wave and tidal cycle for a decade is daunting, to say the least. Because of this, coastal morphological modelling approaches need to rely on some level of process aggregation. High levels of process aggregation are employed in the "behaviour-oriented" modelling approaches described by Stive and de Vriend, 1995 and Stive and Wang, 2003 wherein the general behaviour of large-scale morphological elements is modelled without attempting to describe the low-level physical processes responsible for the changes. The "process-based" modelling approach described in this thesis employs a much lower level of process aggregation. Individual sediment grains and/or waves are not directly modelled, however an attempt is made to explicitly represent all the important physical processes acting on sediment in the coastal environment.

Process-based numerical morphodynamic models are nothing more than attempts by coastal engineers and scientists to distil empirical observations into mathematical formulations relating observed morphological changes to intrinsic properties such as sediment grain size and forcing processes such as tide, waves and wind. The last century has seen the development of these models from simple analytical models to one-dimensional network models, coastal profile models, coast line models, and multi-line models. Each of these approaches simplifies the modelling problem by using formulations which already integrate in two dimensions, leaving only the third dimension free to respond. This approach simplifies the mathematics and reduces the detail required to be captured by the model formulations as a lot of "averaging" is performed in the process of integration. For example, simple one-line coastline models use formulations for predicting total "longshore sediment transport" rates with no regard for where in the cross-shore or vertical dimensions the sediment transport might be occurring. They then predict changes in coastline position by temporally integrating longshore gradients in longshore sediment transport under the assumption of conservation of sediment mass.

During the 1980s and 1990s depth-averaged two-dimensional (2DH) models were developed in which only the vertical dimension was integrated and two-dimensional spatial patterns of morphological change could be described. Originating mainly in river engineering (e.g. Struiksma, 1985) the models often had sophisticated quasi-three-dimensional (quasi-3D) extensions to allow for spiral flow in river bends. Later they were used in coastal areas where waves also play a crucial role in driving currents and sediment transport. For reviews of several such models see de Vriend et al (1993) and Nicholson et al. (1997). Quasi-3D concepts have also been implemented to account for cross-shore processes such as return flow (undertow), bed slope effects, and wave asymmetry effects (e.g. Watanabe et al., 1986; Johnson et al., 1994; Wang et al., 1995; Bos et al., 1996). In coastal environments, such efforts were not entirely successful. The resulting transport fields were extremely sensitive to small disturbances resulting in very irregular patterns of predicted morphological change.

More recently, Péchon and Teisson (1996) presented a morphological model based on a three-dimensional (3D) flow description, where the near-bed velocity was coupled with a local sediment transport formula. This model also produced rather irregular results, which were at least partly due to the assumption of local equilibrium transport. Gessler et al. (1999) developed a 3D model for river morphology, which includes separate equations for bed load transport and 3D suspended transport. It considers several size fractions of sediment and keeps track of the bed composition and evolution during each time step.

As a result, the state-of-the-art at the outset of the research described in this thesis was that quasi-3D and 3D approaches were successfully applied in river engineering, but their practical use in estuarine and coastal applications was limited. The use of straightforward 2DH morphological models had become more or less commonplace, especially in relatively large-scale applications in tidal inlets and estuaries. Ongoing increases in the computing power available to coastal engineers, and the development of techniques to efficiently accelerate the morphological predictions of process-based morphological models meant that morphological simulations of years to decades had become feasible (e.g. Steijn et al., 1998; Roelvink et al., 2001).

However, in complex situations, several fully three-dimensional processes contribute to vertical deviations of the velocity profile from the commonly assumed logarithmic one, such as channel curvature, current acceleration and deceleration, wind- and wave-driven currents, density gradients, and Coriolis force. Also, in locations of local erosion and deposition, the shape of vertical sediment concentration profiles may differ substantially from those found under equilibrium conditions. Conditions such as these are common near the mouths of rivers, in complex estuarine geometries, and near structures. See, for example, Elias (2006) who demonstrates the important impact of even relatively small fresh water discharges on residual flow and sediment transport between the outer coast and the Wadden Sea through the Texel tidal inlet in the north of the Netherlands. Similarly, Zyserman and Johnson (2002) and Lesser et al. (2003) demonstrate the importance of the three-dimensional flow and sediment transport occurring around surface piercing and submerged shore-parallel breakwaters. Predicting the behaviour of such complex systems requires the use of numerical models or model systems that are able to simulate arbitrary combinations of three-dimensional processes within broad classes of problems.

1.2   Objectives

The overarching aim of this thesis is to develop and test improved methods and modelling approaches for the analysis and prediction of coastal morphology on spatial scales of hundreds of metres to kilometres and time scales of several years. Specific objectives are:

1. To develop a new generally applicable coastal morphological model capable of representing the dominant coastal sediment transport processes in a fully three-dimensional manner and to simulate the resulting morphological change occurring over a period of several years. The model should also be simpler to apply and more robust than existing coastal morphological models.

2. To validate the individual process models contained within this new model against available theories, laboratory, and field data.

3. To apply the total morphological model in a complex coastal environment with a well-described morphological change signal and well known forcing conditions, and

4. To isolate and quantify the errors introduced by use of the morphological acceleration methods used with the new model and to compare the significance of these errors with other errors present in the modelling process.

1.3   Thesis Structure and Approach

The structure of this thesis reflects the objectives outlined above. Chapter 2 describes the development of a new three-dimensional coastal morphological model. The model is developed by adding sediment transport and morphological updating formulations to the existing hydrodynamic (FLOW) module of the Delft3D modelling framework developed and marketed by WL|Delft Hydraulics. Delft3D-FLOW is a powerful and robust hydrodynamic engine which has been developed and rigorously tested in a wide range of coastal and non-coastal environments over more than 20 years. It is continuously developed by WL|Delft Hydraulics

and was already capable of modelling most of the important hydrodynamic processes occurring in complex (three-dimensional) coastal zones. The developments
described in Chapter 2 were initially made in a research version of the model
however, once they were sufficiently tested and validated, they were incorporated
into the commercial version of the Delft3D-FLOW module and have been since
been used by researchers and consultants in many regions of the world. Chapter 3
describes validation tests that were devised and performed to test the successful
implementation of the new sediment transport and morphological change formulations in Delft3D-FLOW. The tests begin by reproducing simple analytical and
theoretical sediment transport problems and move on to reproducing more complex laboratory-scale physical model results. The model is also validated against
the results of prototype-scale measurements and previous model results. Chapter 4 reports the application of the model to a real-life coastal erosion problem
occurring at the dynamic entrance of Willapa Bay in the state of Washington,
USA. Willapa Bay is an ideal location for validating coastal morphological models
as it is dynamic, relatively free of human intervention, and well described in terms
of physical process measurements and morphological change. In Chapter 5 the
techniques used to accelerate the morphological simulations of Willapa Bay are
investigated in more detail. Accelerated simulations are compared with benchmark brute-force simulations of five years of morphological change in order to
isolate and quantify the impact of the simplifications and assumptions required to
accelerate morphological simulations using the new Delft3D-FLOW morphological model. Several improvements to existing acceleration techniques are made and
others are identified as worthy of further investigation. Chapter 6 summarises the
key results and findings arrived at during this study and identifies areas requiring
further research.

Chapter 2

A Three-dimensional Morphological Model

Much of the material on which this chapter is based has been previously published in Lesser, G.R., Roelvink, J.A., van Kester, J.A.T.M., Stelling, G.S., 2004. Development and validation of a three-dimensional morphological model. Coastal Engineering 51 *(8-9), 883-915.*

2.1 Introduction

The design of process-based numerical morphological models follows a reasonably standard pattern. Flow velocity fields from a hydrodynamic solver are stored and then used in a sediment transport module to produce sediment transport fields. In the case of coastal morphological models a separate wave module is also required to generate wave-driven currents in the hydrodynamic model and to provide additional sediment stirring in the sediment transport module. Sediment transport fields are often written to a file and accumulated (over a tidal cycle, for example) before being temporally integrated and converted to bed elevation changes in a separate morphological module. Extrapolation or acceleration of the bottom changes computed for a series of sediment transport fields may also be performed by the morphological module. Examples of such models are described by Roelvink and van Banning (1994) and Cayocca (2001) among many others. Advantages of this approach include the clear separation of the various modules which allows removal and replacement of individual modules when improved formulations or approaches are devised and clear inspection of the data passed from one module to another. Disadvantages include the inability to feed back certain information between the modules without incurring computationally expensive iterations of module calls and the need for a complex "steering" module to control the use of the individual modules. Because of the computational expense of passing data from one module to another, updated bathymetry is usually only provided to the wave and hydrodynamic modules relatively infrequently (once every several tides for example).

In this chapter a somewhat different approach is followed. Sediment transport and morphological updating formulations are added directly into an existing hydrodynamic solver, thereby closely coupling the hydrodynamic, sediment transport, and morphodynamic computations. Sediment transport algorithms, predominantly based on the formulations of van Rijn (1993), are added to the Delft3D-FLOW hydrodynamic solver which is widely used, well tested, and well suited to modelling the three-dimensional hydrodynamics of coastal regions. The Delft3D package, developed by WL|Delft Hydraulics in close cooperation with Delft University of Technology, is a model system that consists of a number of integrated modules which together allow the simulation of hydrodynamic flow (under the shallow water assumption), computation of the transport of water-borne constituents (e.g. salinity and heat), short wave generation and propagation, sediment transport and morphological changes, and the modelling of ecological processes and water quality parameters.

At the heart of the Delft3D modelling framework is the FLOW module that performs the hydrodynamic computations and simultaneous (or "online") calculation of the transport of salinity and heat. This chapter describes the addition of online computation of sediment transport and morphological changes within the Delft3D-FLOW module. The main advantages of this online approach are: 1) three-dimensional hydrodynamic processes and the adaptation of non-equilibrium sediment concentration profiles are automatically accounted for in the suspended sediment calculations, 2) the density effects of sediment in suspension (which may cause density currents and/or turbulence damping) are automatically included in the hydrodynamic calculations, 3) changes in bathymetry can be immediately fed back to the hydrodynamic calculations, and 4) sediment transport and morphological simulations are simple to perform and do not require a large data file to communicate results between the hydrodynamic, sediment transport, and bottom updating modules.

The large number of processes already included in Delft3D-FLOW (wind shear, wave forces, tidal forces, density driven flows and stratification due to salinity and/or temperature gradients, atmospheric pressure changes, drying and flooding of inter-tidal flats, etc.) mean that Delft3D-FLOW can be applied to a wide range of river, estuarine, and coastal environments. The online sediment version allows calculation of morphological changes due to the transport, erosion, and deposition of both cohesive (mud) and non-cohesive (sand) sediments in conjunction with any combination of the above processes. This makes the online sediment version of Delft3D-FLOW especially useful for investigating sedimentation and erosion problems in complex hydrodynamic situations.

This chapter describes the physical formulations and numerical implementation used to model the transport of non-cohesive (sand) sediment within Delft3D-FLOW. The work described in this chapter is a continuation of the work described by Lesser (2000) which detailed the implementation of suspended sediment formulations in Delft3D-FLOW. Both cohesive and non-cohesive formulations were implemented, however only the non-cohesive formulations are utilised in this thesis. The Delft3D package is under continual development and a thorough description of the use and extension of the cohesive sediment transport formulations implemented by Lesser (2000) can be found in Ledden (2001). Further development work by other researchers, not described in this thesis, is also ongoing with the addition of multiple sediment fractions, bed stratigraphy, and fluid mud layers all taking place since the developments described in this chapter were implemented.

The developments described in this chapter must be seen as a single step in a long journey. The work described here builds upon the work of many others responsible for the earlier development of the Delft3D-FLOW model and has itself since been built upon by other researchers. Successful coastal morphological models are a moving target, no sooner are the formulations tested and described than they are modified, revised, or improved. This chapter describes the approach and many of the details required to model online sediment transport and morphology in the Delft3D-FLOW module. The resulting model is used for the simulations described throughout the remainder of this thesis.

2.2   Hydrodynamics

2.2.1   Governing equations

The Delft3D-FLOW module solves the unsteady shallow-water equations in two (depth-averaged) or three dimensions. The system of equations consists of the horizontal momentum equations, the continuity equation, the transport equation, and a turbulence closure model. The vertical momentum equation is reduced to the hydrostatic pressure relation as vertical accelerations are assumed to be small compared to gravitational acceleration and are not taken into account. This makes the Delft3D-FLOW model suitable for modelling hydrodynamics in shallow seas, coastal areas, estuaries, lagoons, rivers, and lakes. It aims to model flow phenomena of which the horizontal length and time scales are significantly larger than the vertical scales.

The user may choose whether to solve the hydrodynamic equations on a Cartesian rectangular, orthogonal curvilinear (boundary fitted), or spherical grid. In three-dimensional simulations a boundary fitted (σ-coordinate) approach is used for the vertical grid direction. For the sake of clarity the equations are presented in their Cartesian rectangular form only.

Vertical σ-coordinate system

The vertical σ-coordinate is scaled as $(-1 \leq \sigma \leq 0)$ where

$$\sigma = \frac{z - \zeta}{h} \tag{2.1}$$

The flow domain of a 3D shallow water model consists of a number of layers. In a σ-coordinate system, the layer interfaces are chosen following planes of constant σ. Thus, the number of layers is constant over the horizontal computational area (Figure 2.1). For each layer a set of coupled conservation equations is solved. The partial derivatives in the original Cartesian coordinate system are expressed in σ-coordinates by use of the chain rule. This introduces additional terms (Stelling and Van Kester, 1994).

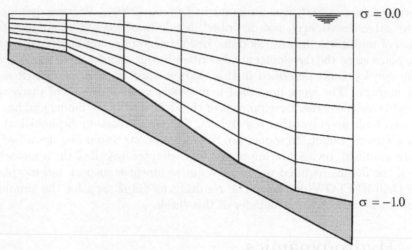

Figure 2.1 – *An example of a vertical grid consisting of six equal thickness σ-layers. Courtesy of Delft Hydraulics.*

Generalised Lagrangian mean (GLM) reference frame

In simulations including waves the hydrodynamic equations are written and solved in a GLM reference frame (Andrews and McIntyre, 1978; Groeneweg and Klopman, 1998; and Groeneweg 1999). In GLM formulation the 2DH and 3D flow equations are very similar to the standard Eulerian equations, however, the wave-induced driving forces averaged over the wave period are more accurately expressed. The relationship between the GLM velocity and the Eulerian velocity is given by:

$$U = u + u_s$$
$$V = v + v_s \tag{2.2}$$

where U and V are GLM velocity components, u and v are Eulerian velocity components, and u_s and v_s are the Stokes' drift components. Walstra et al. (2000) discuss details and verification results.

Hydrostatic pressure assumption

Under the so-called 'shallow water assumption' the vertical momentum equation reduces to the hydrostatic pressure equation. Under this assumption vertical accelerations due to buoyancy effects or sudden variations in the bottom topography are assumed negligible compared to gravitational acceleration and are not taken into account. The resulting expression is:

$$\frac{\partial P}{\partial \sigma} = -\rho g \, h \tag{2.3}$$

Horizontal momentum equations

The horizontal momentum equations are

$$\frac{\partial U}{\partial t} + U\frac{\partial U}{\partial x} + V\frac{\partial U}{\partial y} + \frac{\omega}{h}\frac{\partial U}{\partial \sigma} - fV = -\frac{1}{\rho_0}P_x + F_x + M_x + \frac{1}{h^2}\frac{\partial}{\partial \sigma}\left(\nu_V \frac{\partial U}{\partial \sigma}\right)$$

$$\frac{\partial V}{\partial t} + U\frac{\partial V}{\partial x} + V\frac{\partial V}{\partial y} + \frac{\omega}{h}\frac{\partial V}{\partial \sigma} + fU = -\frac{1}{\rho_0}P_y + F_y + M_y + \frac{1}{h^2}\frac{\partial}{\partial \sigma}\left(\nu_V \frac{\partial V}{\partial \sigma}\right)$$

(2.4)

in which the horizontal pressure terms, P_x and P_y, are given by (Boussinesq approximations)

$$\frac{1}{\rho_0}P_x = g\frac{\partial \zeta}{\partial x} + g\frac{h}{\rho_0}\int_\sigma^0 \left(\frac{\partial \rho}{\partial x} + \frac{\partial \sigma'}{\partial x}\frac{\partial \rho}{\partial \sigma'}\right)d\sigma'$$

$$\frac{1}{\rho_0}P_y = g\frac{\partial \zeta}{\partial y} + g\frac{h}{\rho_0}\int_\sigma^0 \left(\frac{\partial \rho}{\partial y} + \frac{\partial \sigma'}{\partial y}\frac{\partial \rho}{\partial \sigma'}\right)d\sigma'$$

(2.5)

The horizontal Reynold's stresses, F_x and F_y, are determined using the eddy viscosity concept (e.g. Rodi, 1984). For large scale simulations (when shear stresses along closed boundaries may be neglected) the forces F_x and F_y reduce to the simplified formulations

$$F_x = \nu_H \left(\frac{\partial^2 U}{\partial x^2} + \frac{\partial^2 U}{\partial y^2}\right) \qquad F_y = \nu_H \left(\frac{\partial^2 V}{\partial x^2} + \frac{\partial^2 V}{\partial y^2}\right) \qquad (2.6)$$

(Blumberg and Mellor, 1985) in which the gradients are taken along σ-planes. In equation 2.4 M_x and M_y represent the contributions due to external sources or sinks of momentum (external forces by hydraulic structures, discharge or withdrawal of water, wave stresses, etc.).

Continuity equation

The depth-averaged continuity equation is given by

$$\frac{\partial \zeta}{\partial t} + \frac{\partial \left[h\bar{U}\right]}{\partial x} + \frac{\partial \left[h\bar{V}\right]}{\partial y} = S \qquad (2.7)$$

in which S represents the contributions per unit area due to the discharge or withdrawal of water, evaporation, and precipitation.

Transport equation

The advection-diffusion equation reads

$$\frac{\partial [hc]}{\partial t} + \frac{\partial [hUc]}{\partial x} + \frac{\partial [hVc]}{\partial y} + \frac{\partial (\omega c)}{\partial \sigma} =$$
$$h\left[\frac{\partial}{\partial x}\left(D_H \frac{\partial c}{\partial x}\right) + \frac{\partial}{\partial y}\left(D_H \frac{\partial c}{\partial y}\right)\right] + \frac{1}{h}\frac{\partial}{\partial \sigma}\left[D_V \frac{\partial c}{\partial \sigma}\right] + hS$$

(2.8)

in which S represents source and sink terms per unit area.

In order to solve these equations the horizontal and vertical viscosity (ν_H and ν_V) and diffusivity (D_H and D_V) need to be prescribed. In Delft3D-FLOW the horizontal viscosity and diffusivity are assumed to be a superposition of three parts: 1) molecular viscosity, 2) '3D turbulence', and 3) '2D turbulence'. The molecular viscosity of the fluid (water) is a constant value $O(10^{-6})$. In a 3D simulation '3D turbulence' is computed by the selected turbulence closure model (see the following turbulence closure model section). '2D turbulence' is a measure of the horizontal mixing that is not resolved by advection on the horizontal computational grid. 2D turbulence values may either be specified by the user as a constant or space-varying parameter, or can be computed using a sub-grid model for horizontal large eddy simulation (HLES). The HLES model available in Delft3D-FLOW is based on theoretical considerations presented by Uittenbogaard (1998) and is fully discussed by van Vossen (2000).

For use in the transport equation, the vertical eddy diffusivity is scaled from the vertical eddy viscosity according to

$$D_V = \frac{\nu_V}{\sigma_c} \tag{2.9}$$

in which σ_c is the Prandtl-Schmidt number given by

$$\sigma_c = \sigma_{c0} F_\sigma \left(Ri \right) \tag{2.10}$$

where σ_{c0} is purely a function of the substance being transported. In the case of the algebraic turbulence model, $F_\sigma \left(Ri \right)$ is a damping function that depends on the amount of density stratification present via the gradient Richardson's number (Simonin et al., 1989). The damping function, $F_\sigma \left(Ri \right)$, is set equal to 1.0 if the $k-\varepsilon$ turbulence model is used, as the buoyancy term in the $k - \varepsilon$ model automatically accounts for turbulence-damping effects caused by vertical density gradients.

Note that the vertical eddy diffusivity used for calculating the transport of 'sand' sediment constituents may, under some circumstances, vary somewhat from that given by equation 2.9 above. The diffusion coefficient used for sand sediment is described in more detail on page 18 of this thesis.

Turbulence closure models

Several turbulence closure models are implemented in Delft3D-FLOW. All models are based on the so-called 'eddy viscosity' concept (Kolmogorov, 1942; Prandtl, 1945). The eddy viscosity in the models has the following form

$$\nu_V = c'_\mu L \sqrt{k} \tag{2.11}$$

in which c'_μ is a constant determined by calibration, L is the mixing length, and k is the turbulent kinetic energy.

Two turbulence closure models are used in the simulations presented in this thesis. The first is the 'algebraic' turbulence closure model that uses algebraic formulas to determine k and L and therefore the vertical eddy viscosity. The second is the $k - \varepsilon$ turbulence closure model in which both the turbulent energy k and dissipation ε are produced by production terms representing shear stresses at the bed, surface, and in the flow. The 'concentrations' of k and ε in every grid cell are then calculated by transport equations. The mixing length L is determined from ε and k according to

$$L = c_D \frac{k\sqrt{k}}{\varepsilon} \tag{2.12}$$

in which c_D is another calibration constant.

2.2.2   Boundary conditions

In order to solve the systems of equations, the following boundary conditions are required:

Bed and free surface boundary conditions

In the σ-coordinate system the bed and the free surface correspond with σ-planes. Therefore the vertical velocities at these boundaries are simply

$$\omega(-1) = 0 \quad and \quad \omega(0) = 0 \tag{2.13}$$

Friction is applied at the bed as follows:

$$\left. \frac{\nu_V}{h} \frac{\partial u}{\partial \sigma} \right|_{\sigma=-1} = \frac{\tau_{bx}}{\rho} \qquad \left. \frac{\nu_V}{h} \frac{\partial v}{\partial \sigma} \right|_{\sigma=-1} = \frac{\tau_{by}}{\rho} \tag{2.14}$$

where τ_{bx} and τ_{by} are bed shear stress components which may include the effects of wave-current interaction.

Friction due to wind stress at the water surface may be included in a similar manner. For the transport boundary conditions the vertical diffusive fluxes through the free surface and bed are set to zero.

Lateral boundary conditions

Along closed boundaries the velocity component perpendicular to the closed boundary is set to zero (a free-slip condition). At open boundaries one of the following types of boundary conditions must be specified: water level, velocity (in the direction normal to the boundary), discharge, or linearised Riemann invariant (weakly reflective boundary condition, Verboom and Slob, 1984). Additionally, in the case of 3D models, the user must prescribe the use of either a uniform or logarithmic velocity profile at inflow boundaries.

For the transport boundary conditions we assume that the horizontal transport of dissolved substances is dominated by advection. This means that at an open inflow boundary a boundary condition is needed. During outflow the concentration must be free. Delft3D-FLOW allows the user to prescribe the concentration at every σ-layer using a time series. For sand sediment fractions the local equilibrium sediment concentration profile may be used.

2.2.3   Solution procedure

Delft3D-FLOW is a numerical model based on finite differences. To discretise the 3D shallow water equations in space, the model area is covered by a rectangular, curvilinear, or spherical grid. It is assumed that the grid is orthogonal and well-structured. The variables are arranged in a pattern called the Arakawa C-grid (a staggered grid). In this arrangement the water level points (pressure points) are defined in the centre of a (continuity) cell; the velocity components are perpendicular to the grid cell faces where they are situated (see Figure 2.2).

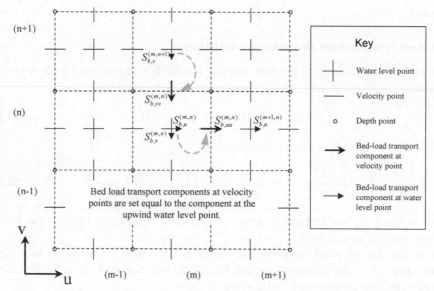

Figure 2.2 – *The Delft3D staggered grid showing the upwind method of setting bed-load sediment transport components at velocity points. Water-level points are located in the centre of the sediment control volumes.*

Hydrodynamics

For the simulations presented in this thesis an alternating direction implicit (ADI) method is used to solve the continuity and horizontal momentum equations (Leendertse, 1987). The advantage of the ADI method is that the implicitly integrated water levels and velocities are coupled along grid lines, leading to systems of equations with a small bandwidth. Stelling extended the ADI method of Leendertse with a special approach for the horizontal advection terms. This approach splits the third-order upwind finite-difference scheme for the first derivative into two second-order consistent discretisations, a central discretisation and an upwind discretisation, which are successively used in both stages of the ADI-scheme. The scheme is denoted as a 'cyclic method' (Stelling and Leendertse, 1991). This leads to a method that is computationally efficient, at least second-order accurate, and stable at Courant numbers of up to approximately 10. The diffusion tensor is redefined in the σ-coordinate system assuming that the horizontal length scale is much larger than the water depth (Mellor and Blumberg, 1985) and that the flow is of boundary-layer type.

The vertical velocity, ω, in the σ-coordinate system is computed from the continuity equation,

$$\frac{\partial \omega}{\partial \sigma} = -\frac{\partial \zeta}{\partial t} - \frac{\partial [hU]}{\partial x} - \frac{\partial [hV]}{\partial y} \tag{2.15}$$

by integrating in the vertical from the bed to a level σ. At the surface the effects of precipitation and evaporation are taken into account. The vertical velocity, ω, is defined at the iso-σ-surfaces. ω is the vertical velocity relative to the moving σ-plane and may be interpreted as the velocity associated with up- or downwelling motions. The vertical velocities in the Cartesian coordinate system can be expressed in the horizontal velocities, water depths, water levels, and vertical coordinate velocities according to:

$$w = \omega + U\left(\sigma\frac{\partial h}{\partial x} + \frac{\partial \zeta}{\partial x}\right) + V\left(\sigma\frac{\partial h}{\partial y} + \frac{\partial \zeta}{\partial y}\right) + \left(\sigma\frac{\partial h}{\partial t} + \frac{\partial \zeta}{\partial t}\right) \tag{2.16}$$

Transport

The transport equation is formulated in a conservative form (finite-volume approximation) and is also solved using the so-called 'cyclic method' (Stelling and Leendertse, 1991). For steep bottom slopes in combination with vertical stratification, horizontal diffusion along σ-planes introduces artificial vertical diffusion (Huang and Spaulding, 1996). Delft3D-FLOW includes an algorithm to approximate horizontal diffusion along z-planes in a σ-coordinate framework (Stelling and Van Kester, 1994). In addition, a horizontal Forester filter (Forester, 1979) based on diffusion along σ-planes is applied to remove any negative concentration values that may occur. The Forester filter is mass conserving and does not inflict significant amplitude losses in sharply peaked solutions.

2.3   Waves

2.3.1   Approach

Wave effects can be included in a Delft3D-FLOW simulation by running the separate Delft3D-WAVE module. A call to the Delft3D-WAVE module must be made prior to running the FLOW module. This will result in a communication file being stored which contains the results of the wave simulation (RMS wave height, peak spectral period, wave direction, mass fluxes, etc) on the same computational grid as is used by the FLOW module. The FLOW module can then read the wave results and include them in flow calculations. Wave simulations may be performed using the 2nd generation wave model HISWA (Holthuijsen et al., 1989) or the 3rd generation SWAN model (Holthuijsen et al., 1993). A significant practical advantage of using the SWAN model is that it can run on the same curvilinear grids as are commonly used for Delft3D-FLOW calculations; this significantly reduces the effort required to prepare combined WAVE and FLOW simulations.

In situations where the water level, bathymetry, or flow velocity field change significantly during a FLOW simulation, it is often desirable to call the WAVE module more than once. The computed wave field can thereby be updated accounting for the changing water depths and flow velocities. This functionality can be achieved by use of either the MORSYS steering module or, more simply, the new 'Online Wave' interface can be used when only alternating calls to the WAVE and FLOW modules are required (as for most coastal morphological simulations using the online sediment transport and morphology discussed here). At each call to the WAVE module the latest bed elevations, water elevations and, if desired, current velocities are transferred from FLOW.

2.3.2   Wave effects

In coastal seas wave action influences morphology in a number of ways. The following processes are presently accounted for in Delft3D-FLOW.

Wave forces

Waves exert forces on the mean hydrodynamic flows. In the nearshore zone these forces are usually dominated by near-surface forces created by the dissipation of wave energy due to wave breaking. In the standard Delft3D model, wave forces due to breaking are modelled as a shear stress at the water surface (Svendsen, 1985; Stive and Wind, 1986). This shear stress can be determined using either the simplified expression of Dingemans et al. (1987), where contributions other than those related to the dissipation of wave energy are neglected, or by taking gradients in the total radiation stress. Dingemans' approach simply relates the wave force to the rate of wave energy dissipation by

$$\vec{M} = \frac{D}{\omega}\vec{k} \qquad (2.17)$$

in which $\vec{M}=$ Force vector (N/m^2), $D=$ Dissipation due to wave breaking (W/m^2), $\omega=$ Angular wave frequency (rad/s), and $\vec{k}=$ Wave number vector (rad/m).

In the standard Delft3D model, if the option to compute the wave force from gradients in total radiation shear stress is selected, the force is still only applied at the water surface and as such is unsuitable for three-dimensional models as spurious vertical circulation of water (reverse undertow) occurs outside the surfzone where wave shoaling dominates wave breaking.

In a three-dimensional model, if the forces produced by both wave breaking and wave shoaling are to be included, a more appropriate vertical distribution of wave forces is required. In the research version of Delft3D discussed here, a simple division of wave forces over the vertical was implemented as follows:

1. The total wave force vector is calculated from the total radiation stress gradients obtained from the SWAN wave model.

2. The wave energy dissipation due to bottom friction is estimated according

to

$$D_{BBL} = min \left(f_{\text{FWFAC}} \frac{0.01 \hat{U}_\delta^3}{\sqrt{\pi}}, D_{TOT} \right) \tag{2.18}$$

where D_{TOT} is the total wave energy dissipation, also obtained from the SWAN wave model, and f_{FWFAC} is a user-specified factor (default 1.0).

3. The wave energy dissipation due to breaking is estimated by

$$D_{SURF} = D_{TOT} - D_{BBL} \tag{2.19}$$

4. The force contribution due to wave energy dissipation from wave breaking is estimated by applying Equation 2.17 to D_{SURF}. This force is then applied to the hydrodynamics in the surface computational layer.

5. The force contribution due to wave energy dissipation due to bottom friction is estimated by applying Equation 2.17 to D_{BBL}. This force (which results in wave "streaming") is then applied to the hydrodynamics across the thickness of the bottom wave boundary layer, as described by Walstra et al. (2000).

6. The force contributions due to wave breaking and dissipation by bottom friction are subtracted from the total wave force vector.

7. The remaining wave force vector is assumed to be a body force (presumably due to wave shoaling) and is applied to the hydrodynamics uniformly over the water column.

Turbulence production

Additional turbulence production due to dissipation in the bottom wave boundary layer and due to wave white capping and breaking at the surface is included as extra production terms in the $k - \varepsilon$ turbulence closure model (Walstra et al., 2000).

Enhanced bed shear stress

The effect of the enhanced bed shear stress on the flow simulation due to increased apparent bed roughness caused by wave orbital motions is accounted for by following the parameterisations of Soulsby et al. (1993). Of the several models available, the simulations presented in this thesis use the wave-current interaction model of Fredsoe (1984).

Mass flux

The wave-induced mass flux is included and is adjusted for the vertically non-uniform Stokes drift (Walstra et al., 2000).

2.3.3   Effective current

Currents affect wave propagation as the waves travel in a moving medium. When waves meet an opposing current they become shorter and steeper. The opposite is true for a following current. The SWAN wave model can read in an ambient current field and compute spectral wave propagation on the spatially varying current. In the standard version of SWAN the current field is assumed to be identical for all wave frequencies. In many situations this is an oversimplification as longer period waves "feel" the current to a greater depth than short period waves. Dingemans (1997) proposed calculating an effective current experienced by each wave frequency as

$$\tilde{U} = \frac{2k}{\sinh 2kh} \int_{-h}^{0} \vec{U}(z) \cosh[2k(z+h)] dz \qquad (2.20)$$

where h is the water depth and k is the frequency-dependent wave number.

Adopting this approach for a spectral wave model communicating with a three-dimensional hydrodynamic model leaves two possibilities. The more correct is to pass the full three-dimensional current field to the wave model and inside the wave model compute the effective current felt by each wave frequency. This approach requires modification of the SWAN wave model and communication of large amounts of three-dimensional current data. Such an approach was tested by Westhuysen and Lesser (2007) and applied to the heads of Port Phillip Bay, Melbourne, Australia where strong currents interact with long ocean swells. In this application Westhuysen and Lesser found little advantage in computing the frequency-dependent effective currents, compared to simply using a single effective current. This result was expected however, given the long swell periods and shallow water depths at their test location. A simpler approach is to use the standard SWAN model, but to compute an effective current at each grid cell based on the peak (or mean) wave period expected at that location. This approach is implemented in the model discussed here. When Delft3D generates the current field to pass to SWAN, Equation 2.20 is applied at each grid cell, using a local k value determined from the peak period of the previously computed wave field. This approach required some modification of the Delft3D-WAVE interface but, as only one current field is passed, allowed the use of the standard SWAN wave model. This modification is not applied when a simple depth-averaged FLOW simulation is performed as in this situation only the depth-averaged current velocity is passed to SWAN.

2.4   Suspended Sediment Transport

2.4.1   General

In the online sediment version of Delft3D-FLOW, sediment is added to the list of constituents that can be computed by the transport solver described in Section 2.2. Up to five sediment fractions may be defined. Each fraction must be classified as 'mud' or 'sand' as different formulations are used for the bed-exchange and settling velocity of these different types of sediment. In addition to the transport solver, several auxiliary formulations are required to fully describe the behaviour of the sediment. The formulations implemented are described below.

Density effects

The equation of state used to calculate the density of the water with varying salinity and temperature (Eckart, 1958) is extended to include the density effect of sediment in suspension as follows,

$$\rho = \rho_w + \sum_{l=1}^{LSED} c_{vol}^{(l)} \left(\rho_s^{(l)} - \rho_w \right) \tag{2.21}$$

in which ρ_w is the density of the water including salinity and temperature effects; $c_{vol}^{(l)}$ is the volumetric concentration of sediment fraction l; $\rho_s^{(l)}$ is the density of solid particles for sediment fraction l; $LSED$ is the total number of sediment fractions.

Settling velocity

The settling velocity of sediment is modelled as a function of concentration and salinity, which may have a significant impact when modelling the transport of high concentrations of very fine cohesive sediments. However, for the sand transport cases considered in this thesis, these effects are insignificant. The settling velocity of a sand sediment fraction is calculated following the method of van Rijn (1993) based on the nominal sediment diameter and the relative density of the sediment particles.

2.4.2   Sediment exchange with the bed

The exchange of sediment with the bed is implemented by way of sediment sources and sinks placed near the bottom of the flow. Separate pairs of sediment source and sink terms are required for each sediment fraction. These are calculated and located differently for mud sediment fractions than for sand sediment fractions.

Sand sediments

For sand sediment fractions the approach of van Rijn (1993) is applied. A reference height a is calculated based on the bed roughness. Notionally, the reference height a is located at the crest elevation of the sand ripples or other dominant roughness height. The sediment source and sink terms are located in the first computational cell that is entirely above the reference height (the reference cell). Cells that fall below the reference cell are assumed to rapidly respond to changes in bed shear stress, and have a sediment concentration equal to the concentration of the reference cell. The sediment concentration at the reference height is calculated using a formula adapted from van Rijn (1984) to include the presence of multiple sediment fractions

$$c_a = f_{SUS}\, \eta\, 0.015\, \rho_s \frac{d_{50}}{a} \frac{T_a^{1.5}}{D_*^{0.3}} \tag{2.22}$$

in which c_a is the mass concentration of the sediment fraction at the reference height a; f_{SUS} is a user-specified calibration parameter (default value 1.0); η is the relative availability of the sediment fraction at the bed (=1.0 for all simulations discussed in this thesis); T_a and D_* are the dimensionless bed shear stress and dimensionless particle diameter (as calculated by van Rijn, 1993) respectively. We note that T_a is computed from the flow velocity in the bottom computational layer by assumption of two logarithmic velocity profiles. First a logarithmic velocity profile based on the *apparent* bed roughness (possibly enhanced by wave orbital motion) is assumed to exist between the between the mid-height of the bottom computational layer and the estimated height of the top of the wave boundary layer (if one exists). Then a velocity profile based on the *actual* (specified) bed roughness is assumed to exist within the wave boundary layer. This allows the computation of u_*, and hence T_a, regardless of the relative heights of the middle of the bottom computational layer, the edge of the wave boundary layer, and the z_0 roughness height.

The sediment source and sink terms are then calculated assuming a linear concentration gradient between the calculated reference concentration at height a and the computed concentration in the reference cell. The source and sink terms are split so that the sediment source can be evaluated explicitly, whereas the sink must be included in the transport equation implicitly (Lesser, 2000). The resulting expressions are

$$\begin{aligned} Source &= c_a \left(\frac{D_v}{\Delta z} \right) \\ Sink &= c_{kmx} \left(\frac{D_v}{\Delta z} + w_s \right) \end{aligned} \qquad (2.23)$$

in which D_V is the vertical diffusion coefficient at the bottom of the reference cell, Δz is the vertical distance from the reference level a to the centre of reference cell, and c_{kmx} is the mass concentration of the sediment fraction in question in the reference cell (solved implicitly).

Mud sediments

For mud sediment fractions the source and sink terms are always located in the bottom computational cell and are computed with the well-known Parthenaides and Krone formulations (Parthenaides, 1965). The implementation and verification of mud sediment fractions is not described in this thesis. Interested readers should refer to Lesser (2000) and van Ledden (2001) for further details.

2.4.3   Vertical diffusion coefficient for sediment

For sand sediments, in situations without waves, σ_{c0} in equation 2.10 is set to 1.0 and van Rijn's β factor (used to describe the difference between fluid and granular diffusion, van Rijn, 1993) is included in equation 2.9. This results in

$$D_V = \beta \frac{\nu_V}{\sigma_c} \qquad (2.24)$$

where

$$\beta = 1 + 2 \left[\frac{w_s}{u_{*,c}} \right]^2 \quad \text{limited to the range} \quad 1.0 \le \beta \le 1.5 \qquad (2.25)$$

in which $u_{*,c}$ is the current-related bed shear velocity. Note that when the $k - \varepsilon$ turbulence closure model is selected, σ_c will also be equal to 1.0 in equation 2.24. For the algebraic turbulence model, σ_c is limited to the role of representing the turbulence damping effects of vertical density gradients. For this reason, van Rijn's ϕ factor (van Rijn, 1984) which also accounts for these effects is not implemented.

For simulations including waves, using the $k - \varepsilon$ turbulence closure model, the approach is similar to that described above, with one exception. As van Rijn's β factor is intended to apply to only the current-related mixing, the β factor applied to the total mixing computed by the $k - \varepsilon$ model is reduced according to the expression

$$\beta_{eff} = 1 + (\beta - 1)\frac{\tau_c}{\tau_w + \tau_c} \tag{2.26}$$

where τ_c is the bed shear stress due to current and τ_w is the bed shear stress due to waves. β_{eff} is then applied in place of β in equation 2.24.

For simulations including waves, using the algebraic turbulence model, a different approach is used for the vertical diffusion coefficient of sand sediment fractions. In this case, the diffusion coefficient is calculated using analytical expressions given by van Rijn (1993) for both the current- and wave-related turbulent mixing. The current-related mixing is calculated using the 'parabolic-constant' distribution recommended by van Rijn

$$\begin{aligned} z < 0.5h \quad & D_{V,c} = \kappa\,\beta\,u_{*,c}\,z(1 - z/h) \\ z \geq 0.5h \quad & D_{V,c} = 0.25\,\kappa\,\beta\,u_{*,c}\,h \end{aligned} \tag{2.27}$$

where $D_{V,c}$ is the vertical sediment diffusion coefficient due to currents, and $u_{*,c}$ is the current-related bed shear velocity. In the lower half of the water column, this expression produces similar turbulent mixing values to those produced by the standard algebraic turbulence closure model in current-only situations.

The wave-related mixing is calculated as:

$$\begin{aligned} z \leq \delta_s \quad & D_{V,w} = D_{V,bed} = 0.004\,D_*\,\delta_s\,\hat{U}_\delta \\ z \geq 0.5h \quad & D_{V,w} = D_{V,max} = 0.035\,hH_s/T_p \\ \delta_s < \quad z < 0.5h \quad & D_{V,w} = D_{V,bed} + [D_{V,max} - D_{V,bed}]\left[\frac{z - \delta_s}{0.5h - \delta_s}\right] \end{aligned} \tag{2.28}$$

where $D_{V,w}$ is the vertical sediment diffusion coefficient due to waves, and δ_s is the thickness of the near-bed sediment mixing layer following van Rijn, and $\hat{U}_\delta$ is the peak wave orbital velocity at the edge of the wave boundary layer.

The total vertical sediment diffusion coefficient is then calculated as

$$D_V = \sqrt{D_{V,c}^2 + D_{V,w}^2} \tag{2.29}$$

where D_V is the vertical sediment diffusion coefficient used in the suspended sediment transport calculations for this sediment fraction.

For all sand sediment fractions, the vertical sediment diffusion coefficient for layers below the reference layer is set to a relatively high value. This ensures that the concentration of cells below the reference cell comes rapidly to equilibrium. The above calculation of the vertical sediment diffusion coefficient is repeated for each sediment fraction.

2.4.4   Suspended sediment correction vector

In the online-sediment version of Delft3D-FLOW, the transport of suspended sediment is computed over the entire water column (from $\sigma = -1$ to $\sigma = 0$). However, for sand sediment fractions, van Rijn regards sediment transported below the reference height, a, as belonging to 'bed-load' sediment transport which is computed separately as it responds almost instantaneously to changing bed shear stress and feels the effects of bed slopes. In order to prevent double counting, the suspended sediment transport computed by the FLOW transport solver below the reference height a is estimated using simple central difference schemes for both advection and diffusion and the result is stored as a 'suspended sediment correction vector' ($S_{cor,uu}$ and $S_{cor,vv}$). The direction of the correction vector is reversed, and gradients in this vector are included in the computed morphological changes as described in Section 2.6.

2.4.5   Depth-averaged implementation

Delft3D FLOW can also be applied in depth-averaged mode. In this case only one computational layer is present and the discharge through this one computational layer is assumed to be equivalent to the depth-integrated discharge in an equivalent three-dimensional model. As vertical profiles of velocity, sediment concentration, and turbulent mixing are not resolved in a depth-averaged model, additional assumptions are required in order to compute bed shear stress and depth-integrated sediment transport. The main assumption made is that of a logarithmic velocity profile of the form

$$U = \frac{u_{*,c}}{\kappa} \ln\left(\frac{z}{z_0}\right) \tag{2.30}$$

which can be computed directly from the depth-averaged velocity

$$U = \left[\frac{\bar{U}}{z_0/h - 1 + \ln(h/z_0)}\right] \ln\left(\frac{z}{z_0}\right) \tag{2.31}$$

The depth-averaged advection diffusion equation simplifies from equation 2.8 to

$$\frac{\partial[h\bar{c}]}{\partial t} + \frac{\partial[h\bar{U}\bar{c}]}{\partial x} + \frac{\partial[h\bar{V}\bar{c}]}{\partial y} = h\left[\frac{\partial}{\partial x}\left(D_H \frac{\partial\bar{c}}{\partial x}\right) + \frac{\partial}{\partial y}\left(D_H \frac{\partial\bar{c}}{\partial y}\right)\right] + hS \tag{2.32}$$

where $\bar{c}$ is the depth-averaged sediment concentration and S is a sediment source term given by

$$S = \frac{\bar{c}_{eq} - \bar{c}}{T_s} \tag{2.33}$$

in which $\bar{c}_{eq}$ is the depth-averaged *equilibrium* sediment concentration and T_s is an adaptation time-scale.

The depth-averaged equilibrium concentration $\bar{c}_{eq}$ is derived from

$$\bar{c}_{eq} = \frac{\vec{S}_{sus,eq}}{\left|\bar{\vec{U}}\right| h} \tag{2.34}$$

where $\vec{S}_{sus,eq}$ is the depth-integrated suspended sediment transport vector under equilibrium (steady and uniform) conditions. In principle, $\vec{S}_{sus,eq}$ may be derived from any suspended sediment transport formulation, although only the formulation of van Rijn (1993) is discussed here. If the formulation supplies a computed suspended sediment transport rate then this can be used for $\vec{S}_{sus,eq}$ directly. On the other hand, if the transport equation is intended for three-dimensional application and supplies a near-bed reference concentration, such as the van Rijn formulations described on page 17, then $\vec{S}_{sus,eq}$ is determined by one-dimensional (vertical) numerical integration of

$$\vec{S}_{sus,eq} = \left| \int \vec{U} c_{eq} dz \right| \tag{2.35}$$

using 20 virtual layers in the vertical. The virtual layer thicknesses have a logarithmic distribution with the lowest layer having a thickness of 0.5% of the water depth. $\vec{U}$ is assumed to be a logarithmic velocity profile given by equation 2.31 and the equilibrium sediment concentration profile is computed by numerical integration of the stationary one-dimensional advection-diffusion equation

$$c_{eq} w_s + D_v \frac{dc_{eq}}{dz} = 0 \tag{2.36}$$

where D_v is the vertical sediment diffusion coefficient presented by van Rijn (1993) and given in equations 2.27, 2.28, and 2.29.

The adaptation time-scale is computed according to the method of Gallappatti (1983) and is a function of water depth, fall velocity, and bed shear velocity

$$T_s = \frac{h}{w_s} T_{sd} \tag{2.37}$$

where the dimensionless adaptation time Tsd is given by

$$T_{sd} = w_* \exp \left[\begin{array}{c} (1.547 - 20.12 u_r) \, w_*^3 + \left(326.832 u_r^{2.2047} - 0.2\right) w_*^2 \\ + \, (0.1385 \ln (u_r) - 6.4061) \, w_* + (0.5467 u_r + 2.1963) \end{array} \right] \tag{2.38}$$

where $u_r = u_{*,c}/\bar{U}$ and $w_* = w_s/u_{*,c}$.

2.5   Bed-load Sediment Transport

Bed-load transport is calculated for all sand sediment fractions following the approach described by van Rijn (1993). This accounts for the near-bed sediment transport occurring below the reference height a described above. First, the magnitude and direction of the bed-load sand transport are computed using one of two formulations presented by van Rijn depending on whether waves are present in the simulation. The computed sediment transport vectors are then relocated from water-level points to velocity points using an 'upwind' computational scheme to ensure numerical stability. Finally, the transport components are adjusted for bed-slope effects.

2.5.1   Basic formulation

Simulations without waves

For simulations without waves, the magnitude of the bed-load transport on a horizontal bed is calculated using a formulation provided by van Rijn (personal communication, June 2000)

$$|S_b| = f_{BED}\, \eta\, 0.5 \rho_s\, d_{50}\, u'_*\, D_*^{-0.3} T \tag{2.39}$$

where $|S_b|$ is the bed-load transport rate (kg/m/s); f_{BED} is a user-specified calibration factor (default value 1.0) which is included to allow users to adjust the overall significance of bed-load sediment transport; η is the relative availability of the sediment fraction in the mixing layer; u'_*, D_*, and T are the effective bed shear velocity, the dimensionless particle diameter, and the dimensionless bed-shear stress (all following van Rijn, 1993) respectively. u'_* and T are based on the computed velocity in the bottom computational layer.

In the absence of waves, the direction of the bed-load transport (on a horizontal bed) is taken to be parallel with the flow in the bottom computational layer. Thus the bed-load vector components are given by

$$S_{b,u} = \frac{u_{b,u}}{|u_b|}\, |S_b|\;, \qquad S_{b,v} = \frac{u_{b,v}}{|u_b|}\, |S_b| \tag{2.40}$$

where $u_{b,u}$, $u_{b,v}$, and $|u_b|$ are the local bottom-layer flow velocity components and magnitude.

Simulations including waves

For simulations including waves, the magnitude and direction of the bed-load transport on a horizontal bed are calculated using an approximation method developed by van Rijn (2001). This method includes an estimate of the effects of wave orbital velocity asymmetry on bed-load sediment transport. The method computes the magnitude of the bed-load transport as

$$|S_b| = \eta\, 0.006 \rho_s\, w_s\, M^{0.5}\, M_e^{0.7} \tag{2.41}$$

where $|S_b|$ = magnitude of bed load transport (kg/m/s), η = relative availability of the sediment fraction in the mixing layer, M = sediment mobility number due to waves and currents, and M_e = excess sediment mobility number. M and M_e are computed as:

$$M = \frac{v_{eff}^2}{(s-1)\, g\, d_{50}} \tag{2.42}$$

$$M_e = \frac{(v_{eff} - v_{cr})^2}{(s-1)\, g\, d_{50}} \tag{2.43}$$

where

$$v_{eff} = \sqrt{v_R^2 + U_{on}^2} \tag{2.44}$$

in which s = relative sediment density ($= \rho_s/\rho$); v_{cr} = critical depth-averaged velocity for initiation of motion (based on a parameterisation of the Shields curve); v_R = magnitude of an equivalent depth-averaged velocity computed from the (Eulerian) velocity in the bottom computational layer, assuming a logarithmic velocity profile; U_{on} = near-bed peak orbital velocity in onshore direction (in the direction on wave propagation) based on the significant wave height.

U_{on} (and U_{off} used below) are the high frequency near-bed orbital velocities due to short waves and are computed using a modification of the method of Isobe and Horikawa (1982). This method is a parameterisation of fifth-order Stokes wave theory and third-order cnoidal wave theory. It can be used over a wide range of wave conditions and takes into account the non-linear effects that occur as waves propagate in shallow water (Grasmeijer and van Rijn, 1998).

The direction of the bed-load transport vector is determined by assuming that it is composed of two parts: 1) a part due to the current ($S_{b,c}$) which acts in the direction of the (Eulerian) near-bed current, and 2) a part due to the waves ($S_{b,w}$) which acts in the direction of wave propagation. The magnitudes of these two parts are determined as follows:

$$|S_{b,c}| = \frac{|S_b|}{\sqrt{1 + r^2 + 2\,|r|\cos\varphi}} \tag{2.45}$$

$$|S_{b,w}| = r\ |S_{b,c}| \tag{2.46}$$

where:

$$r = \frac{(|U_{on}| - v_{cr})^3}{(|v_R| - v_{cr})^3} \tag{2.47}$$

$S_{b,w} = 0$ if $r < 0.01$, $S_{b,c} = 0$ if $r > 100$, and φ = angle between current and wave directions.

Also included in the "bed-load" transport vector is an estimation of the *suspended* sediment transport due to wave asymmetry effects. This is intended to model the effect of asymmetric wave orbital velocities on the transport of suspended material within approximately 0.5 m of the bed and accounts for the bulk of the suspended transport affected by high frequency wave oscillations.

This wave-related suspended sediment transport is modelled using an approximation method proposed by van Rijn (2001):

$$|S_{s,w}| = \gamma\, U_A\, L_T \tag{2.48}$$

where: $|S_{s,w}|$ = magnitude of the wave-related suspended transport (kg/m/s), γ = phase lag coefficient ($= 0.2$), U_A = velocity asymmetry value $= \frac{U_{on}^4 - U_{off}^4}{U_{on}^3 + U_{off}^3}$, and L_T = suspended sediment load $= 0.007\,\rho_s\,d_{50}\,M$.

The three separate transport modes are then combined under the assumption that $S_{b,c}$ is in the direction of the (Eulerian) near-bed current and $S_{b,w}$ and $S_{s,w}$ are in the direction of wave propagation. This results in the following bed-load transport components:

$$\begin{aligned}
S_{b,u} &= f_{BED}\left[\frac{u_b}{|\vec{u}_b|}\,|S_{b,c}| + (f_{BEDW}\,S_{b,w} + f_{SUSW}\,S_{s,w})\cos\phi\right]\\
S_{b,v} &= f_{BED}\left[\frac{v_b}{|\vec{u}_b|}\,|S_{b,c}| + (f_{BEDW}\,S_{b,w} + f_{SUSW}\,S_{s,w})\sin\phi\right]
\end{aligned} \tag{2.49}$$

where f_{BED} = user-specified calibration factor (default value 1.0), f_{BEDW} = user-specified calibration factor (default value 1.0), f_{SUSW} = user-specified calibration factor (0.5 recommended for field cases, 1.0 for flumes), $u_b, v_b, \vec{u}_b$ = Eularian velocity components and vector in the bottom computational layer, and ϕ = local angle between the direction of wave propagation and the computational grid.

2.5.2   Upwind shift

The bed-load transport vector components described above are computed at the water-level points in the Delft3D-FLOW staggered grid (e.g. $S_{b,u}$ in Figure 2.2), as are the suspended-sediment sources and sinks. The bed-load vector components at the velocity points (e.g. $S_{b,uu}$), around the perimeter of each cell control volume, are determined by transferring the appropriate vector components from the adjacent water-level point half a grid cell 'upwind', as indicated in Figure 2.2. The upwind direction is based on the computed direction of the bed-load transport vectors in the water-level points. If the vector directions in adjacent water-level points oppose then a simple vector mean is used. This upwind shift ensures numerical stability and allows the implementation of an extremely simple morphological updating scheme, as described in Section 2.6.

In the example shown in Figure 2.2 the bed-load sediment transport components at the u- and v-velocity points of grid cell (m,n) are set as follows: In the u direction, the transport at the u-velocity point, $S_{b,uu}^{(m,\,n)}$, is set equal to the u-component of the transport computed at the upwind water-level point, in this case, $S_{b,u}^{(m,\,n)}$. In the v direction, the transport at the v-velocity point, $S_{b,vv}^{(m,\,n)}$, is in this case set equal to $S_{b,v}^{(m,\,n+1)}$ because the bed-load transport direction opposes the grid direction. Following the upwind shift, the bed-load transports at the U and V velocity points are then modified for bed-slope effects.

2.5.3   Bed slope effects

As bed-load transport is more-or-less continuously in contact with the bed, the slope of the bed affects the magnitude and direction of the bed-load transport vector. A longitudinal slope in the direction of the bed-load transport modifies the magnitude of the bed-load vector as follows (modified from Bagnold, 1966)

$$S_{b,uu} = \alpha_s S_{b,uu}\,, \qquad S_{b,vv} = \alpha_s\, S_{b,vv} \qquad (2.50)$$

where

$$\alpha_s = 1 + f_{ALFABS} \left[\frac{\tan\left(\phi\right)}{\cos\left(\tan^{-1}\left(\frac{\partial z}{\partial s}\right)\right)\left(\tan\left(\phi\right) - \frac{\partial z}{\partial s}\right)} - 1 \right] \qquad (2.51)$$

in which f_{ALFABS} is a user-specified tuning parameter; $\partial z/\partial s$ is the bed slope in the direction of the bed-load transport (positive down); ϕ is the internal angle of friction of bed material (assumed to be 30°).

A transverse bed slope modifies the direction of the bed-load transport vector. This modification is broadly based on the work of Ikeda (1982) and is computed as follows:

$$S_{b,uu} = S_{b,uu} - \alpha_n \, S_{b,vv} \,, \qquad\qquad S_{b,vv} = S_{b,vv} + \alpha_n \, S_{b,uu} \qquad (2.52)$$

where

$$\alpha_n = f_{ALFABN} \left(\frac{\tau_{b,cr}}{\tau_{b,cw}} \right)^{0.5} \frac{\partial z}{\partial n} \qquad (2.53)$$

in which f_{ALFABN} is a user-specified coefficient (default 1.5); $\tau_{b,cr}$ is the critical bed shear stress; $\tau_{b,cw}$ is the bed shear stress due to current and waves; $\partial z/\partial n$ is the bed slope normal to the unadjusted bed-load transport vector.

2.6 Morphodynamics

The quantity of each sediment fraction available at the bed is computed every half time step using simple bookkeeping for the control volume of each computational cell. This simple approach is made possible by the upwind shift of the bed-load transport components described above and the very frequent feedback from the morphological change to the hydrodynamics made possible by the online sediment implementation. This method is much less computationally intensive than the Lax-Wendroff bed updating scheme used in many other morphological models.

2.6.1 Sediment fluxes

Suspended sediment transport

The net sediment change due to suspended sediment transport is calculated as follows

$$\Delta s_{\text{sus}}^{(m,n)} = f_{\text{MORFAC}} \left(Sink - Source \right) \Delta t \qquad (2.54)$$

where f_{MORFAC} is the morphological acceleration factor (described on Page 26); $Sink$ and $Source$ are the suspended-sediment sink and source terms as given by equation 2.23 above; and Δt is the computational (half) time step.

The correction for suspended sediment transported below the reference height, a, is taken into account by including gradients in the suspended transport correction vector, S_{cor}, as follows:

$$\Delta s_{\text{cor}}^{(m,n)} = f_{\text{MORFAC}} \left(\begin{array}{c} S_{cor,uu}^{(m-1,\,n)} \, \Delta y^{(m-1,\,n)} - S_{cor,uu}^{(m,\,n)} \, \Delta y^{(m,n)} + \\ S_{cor,vv}^{(m,\,n-1)} \, \Delta x^{(m,\,n-1)} - S_{cor,vv}^{(m,\,n)} \, \Delta x^{(m,n)} \end{array} \right) \frac{\Delta t}{A^{(m,n)}}$$

$$(2.55)$$

where: $A^{(m,n)}$ is the area of computational cell at location (m,n); $S_{cor,uu}^{(m,\,n)}$ and $S_{cor,vv}^{(m,\,n)}$ are the suspended-sediment correction vector components in the u and v directions, at the u and v velocity points of the computational cell at location (m,n); $\Delta x^{(m,\,n)}$ and $\Delta y^{(m,\,n)}$ are the widths of cell (m,n) in the x and y directions respectively.

Bed-load sediment transport

Similarly, the change in bottom sediment due to bed-load transport is calculated as

$$\Delta s_{\text{bed}}^{(m,n)} = f_{\text{MORFAC}} \left(\begin{array}{l} S_{b,uu}^{(m-1,n)} \Delta y^{(m-1,n)} - S_{b,uu}^{(m,n)} \Delta y^{(m,n)} + \\ S_{b,vv}^{(m,n-1)} \Delta x^{(m,n-1)} - S_{b,vv}^{(m,n)} \Delta x^{(m,n)} \end{array} \right) \frac{\Delta t}{A^{(m,n)}}$$

(2.56)

where $S_{b,uu}^{(m,n)}$ and $S_{b,vv}^{(m,n)}$ are the bed-load sediment transport vector components at the u and v velocity points respectively.

Total change in bed sediments

The total change in sediment is simply the sum of the change due to suspended load, the change due to the suspended-load correction vector, and the change due to bed load. This process is repeated for each sediment fraction.

2.6.2   Morphological acceleration factor

The morphological acceleration factor (morfac, f_{MORFAC}) is a device used to assist in dealing with the difference in time-scales between hydrodynamic and morphological developments. It works very simply by multiplying the sediment fluxes to and from the bed by a constant factor, see equations 2.54, 2.55, 2.56, thereby effectively extending the morphological time step. Effectively

$$\Delta t_{\text{morphology}} = f_{\text{MORFAC}} \, \Delta t_{\text{hydrodynamic}}$$

(2.57)

This technique allows long morphological simulations to be achieved using hydrodynamic simulations of only a fraction of the required duration. Obviously, there are limits to the morphological acceleration factor that can be applied, depending on the characteristics of the location under consideration. The selection of a suitable morphological acceleration factor remains a matter of judgement and sensitivity testing for the modeller. Several test cases applying different morphological acceleration factors have been performed during the validation process and are presented in Chapter 3. The application of a morphological factor in coastal environments is discussed further in Chapter 5.

2.6.3   Fixed layers

Expressions are also included which limit the erosion due to gradients in suspended and bed-load sediment transport if the quantity of sediment at the bed approaches zero (i.e. a fixed layer is approached). This is achieved by multiplying both the bed-load transport components and the suspended sediment sources and sinks by a reduction factor f_{FIXFAC} if the available depth of sediment at the bed falls below a user-specified threshold $Thresh$, as follows:

$$f_{\text{FIXFAC}} = \min \left(\frac{\Delta_{sed}}{Thresh}, 1 \right)$$

(2.58)

where Δ_{sed} is the thickness of sediment available at the bed. This reduction in the source and sink terms is only applied if erosive conditions are expected. To assess this, the unadjusted source and sink terms are evaluated using the sediment concentration from the previous time step. If depositional conditions are expected then f_{FIXFAC} is set to 1.0 so that deposition close to a fixed layer is not hindered. Bed load-sediment transport vector components at velocity points in the Delft3D staggered grid are modified by multiplying by the f_{FIXFAC} calculated at the *upwind* water-level point. This effectively limits the quantity of bed-load sediment that can *leave* a grid cell.

This approach conserves sediment mass and has been proven to be robust and stable. It does however have two known issues: 1) it will result in artificially low concentrations of sediment in suspension if $f_{MORFAC} > 1.0$ and erosion is limited by a fixed layer and 2) it requires a depth of bed sediment at least equal to *Thresh* for bed-load sediment fluxes to propagate across a grid cell unhindered. The former limitation has implications for comparison of measured and modelled sediment concentrations and possibly for situations where sediment concentration plays a critical role in density-driven stratification or currents. The latter limitation implies that grid cells that are initialised completely bare of sediment will be forced to accumulate a little sediment before any bed-load transport will be able to cross them. If large areas of the model are bare of sediment they should perhaps be initialised with a little sediment present to prevent unrealistic accumulation of bed-load sediment in these grid cells.

2.6.4   Erosion of dry cells

One of the problems that may affect traditional numerical morphological models is that of bank erosion. This problem usually occurs because 'wet' or computationally active cells can erode however 'dry' cells do not experience any change in sediment mass or bed elevation and can only be brought into the active domain if the water level rises sufficiently to cause inundation.

Various techniques have been devised to address this problem in the literature (e.g. Roelvink et al., 2006b). One of the fundamental problems faced in implementing any solution is that the bank erosion (slumping etc.) processes that occur in nature due to the relatively steep slope of the banks cannot be adequately represented on the relatively coarse computational grid required for useful duration morphological simulations. One of the more promising techniques is presented by Olsen (2003) who implements an adaptive model grid with fine grid cells located near the edges of the active computational domain to better resolve bank erosion processes.

As part of this study, a rather simple approach to the bank erosion problem was devised and implemented in the online sediment version of Delft3D-FLOW. It is described as "dry-wet cell bank erosion" in Roelvink et al. (2006b) and works by simply (partially) transferring erosion occurring in the last "wet" grid cell to any adjacent "dry" grid cells. The method requires the user to specify two parameters: Θ_{SD}, the maximum fraction of erosion to reallocate from edge wet cells to surrounding dry cell(s), and $h_{max\Theta}$, the water depth in the wet cell at which the full Θ_{SD} will be reallocated. The actual fraction of erosion in an edge wet cell that is reallocated is determined by

$$\Theta = \min\left(\frac{h - h_{sedThr}}{h_{max\Theta} - h_{sedThr}}, 1\right)\Theta_{SD} \qquad (2.59)$$

where h_{sedThr} is the minimum threshold depth for performing sediment calculations.

There are two possible methods of configuring this approach to bank erosion. The first uses a high Θ_{SD} value (e.g. $\Theta_{SD} = 1.0$ and a positive value for $h_{max\Theta}$. As the edge wet grid cell erodes an increasing fraction of the erosion is applied to the surrounding dry grid cells. Once the depth in the wet grid cell reaches $h_{max\Theta}$ then all further erosion is applied to the adjacent dry grid cell(s) and the edge wet grid cell will not erode further until all surrounding grid cells are also wet. The alternative approach is to set a more moderate value for Θ_{SD} and a near-zero value for $h_{max\Theta}$. In this way the fraction Θ_{SD} of the erosion of edge wet cells will simply be reallocated to adjacent dry cells, until the adjacent dry cells all become wet. Experience has shown that the latter method is more effective at eroding a dry beach.

Physically, this approach amounts to assuming that the initial profile shape between two adjacent grid cells can only vary by a small (user-specified) amount before the landward grid cell is also brought down. The approach is admittedly crude, however it does achieve the key objective of creating a land-side supply of sediment for an eroding (beach) profile. Tests by Roelvink et al. (2003) indicate that the method can achieve realistic results compared to experimental data of breach growth. Experience with modelling eroding beach profiles (e.g. in the offshore breakwater validation case discussed in Section 3.4.2 confirms that activating this dry bank erosion option keeps profiles much more realistic and prevents simulations from becoming unstable due to unrealistically high beach slopes near the water line. The method has also been successfully applied to modelling 800 years of development of meandering tidal channels by van der Wegan et al. (2006).

2.6.5   Feedback to hydrodynamics

The depth to the bed in water-level and velocity points is (optionally) updated every half time step. This takes into account the total change in mass of all sediment fractions present in a computational cell. For the simulations discussed in this thesis sediment fractions are assumed to be independent and instantly mixed. This simple bottom model is perfectly adequate for simulations which only use one sediment fraction. An improved bed model is the subject of ongoing development efforts and more advanced bottom updating models are available in research versions of the code (e.g. as implemented by van Ledden and Wang, 2001).

In order to ensure stability of the morphological updating procedure it is important to ensure a one-to-one coupling between bottom elevation changes and changes in the bed shear stress used for bed-load transport and sediment source and sink terms. Because of the staggered grid used by Delft3D this requires the use of a combination of numerical schemes, as follows:

- Depth in water-level points is updated on the basis of the changed mass of sediment in each control volume.

- Depth in velocity points is set equal to the minimum of the depths in adjacent water level points.

- Bed shear stress in water-level points (used for computing bed-load sediment transport and suspended sediment source and sink terms) is determined by averaging "discharge" at adjacent velocity points and recomputing local velocity by dividing by the water depth at the water-level point. In this context, "discharge" at adjacent velocity points is computed using the velocity in the lowest computational layer and the corresponding water depth in the velocity point.

- Bed load transport applied at velocity points is taken from upwind water-level points.

2.7   Summary

Sediment transport and morphological change formulations have been integrated into the existing powerful three-dimensional hydrodynamic model Delft3D-FLOW which is capable of representing most coastal processes including many of the three-dimensional effects of waves.

The new tightly integrated ("online") sediment transport formulations give Delft3D-FLOW the ability to model the erosion, transport, and deposition of both cohesive and non-cohesive sediment fractions. Both 2DH and 3D simulations are possible, although the 2DH implementation relies on additional assumptions regarding velocity profiles, turbulent mixing profiles, and sediment adaptation time-scales, all of which are explicitly modelled in 3D simulations.

Suspended and bed-load sediment transport modes are separately accounted for. Bed-load transport is assumed to react instantly to changes in bottom shear stress and to feel the effects of bed slopes. Suspended sediment is introduced into the flow by means of sediment sources and sinks placed near the bed. In three-dimensional simulations suspended sediment transport rates will adapt over natural time and length scales to changes in bed shear stress. In two-dimensional (depth averaged) simulations the adaptation scales are determined using an empirical approach.

The tight integration of the online sediment transport has several advantages over previous methods:

1. It is simpler to construct process-based morphological models as a complex steering module is no longer required.

2. It allows simple treatment of 3D and non-equilibrium transport and feedback of sediment density effects to the hydrodynamics.

3. It has enabled the implementation of a new morphological acceleration technique using a simple morphological acceleration factor (morfac) which is

applied to sediment fluxes to and from the bed. Initial testing indicates the method is robust, accurate, and simple to apply.

4. Very frequent feedback of morphological changes to the hydrodynamics help to keep the morphological simulation stable and eliminate the need for a complex bed updating scheme.

5. The tight coupling also allowed simple approaches to fixed layers and bank erosion to be implemented. Initial testing indicates the methods are robust and qualitatively reasonable.

The three-dimensional effects of waves on flow have been improved by separating wave forces into near-bed, surface, and body forces. The Delft3D-WAVE interface has also been improved so that three-dimensional hydrodynamic simulations provide a depth-weighted current to the SWAN wave model. The weighting assigned to currents at various depths depends on the local peak wave period.

The vertical mixing of sediment in three-dimensional simulations is either from the built-in $k - \varepsilon$ turbulence model, which is modified to include extra turbulence generation by waves, or by analytical expressions given by van Rijn (1993).

Chapter 3
Model Validation

Much of the material on which this chapter is based has been previously published in Lesser, G.R., Roelvink, J.A., van Kester, J.A.T.M., Stelling, G.S., 2004. Development and validation of a three-dimensional morphological model. Coastal Engineering 51 *(8-9), 883-915.*

3.1 Introduction

One of the advantages of process-based numerical morphological models is that they can be applied to reproducing morphological development at a wide range of scales. From 10cm deep flows in laboratory flumes to 100m deep channels in major estuaries the physical processes governing hydrodynamics and sediment transport hardly vary. As no scale factors need to be applied to numerical simulations, unlike physical morphological models where the scaling of sediment is very problematic, numerical morphological models can be directly tested against both laboratory- and prototype-scale observations. This considerably simplifies the challenge of validating a process-based morphological model as the individual process formulations can be validated one by one or in small groups of closely related processes by careful selection of validation test cases.

Validation of Deflt3D-FLOW with online morphology begins with four simple test cases for which analytical solutions exist. These tests demonstrate the development of the expected equilibrium sediment concentration profile under stationary conditions and confirm the correct numerical implementation of the sediment pick-up, settling, and morphological development formulations. They also serve to indicate the required vertical grid resolution. Following this, three tests against laboratory flume experiments are presented. The effects of flow deceleration and acceleration are tested against a flume test of flow over a steep-sided trench conducted by van Rijn (1987). Flow curvature and bed slope effects are tested against a curved flume experiment by Struiksma et al. (1984), and the combined effect of waves and currents on equilibrium sediment concentration profiles is tested against flume experiments conducted by Dekker and Jacobs (2000). The final test for which real-world validation data exists is a prototype-scale simulation of the morphological development that occurred around the extended breakwaters at the Dutch port of IJmuiden. This is the only tide-dominated simulation presented in this chapter and is used to test: 1) the "morfac" morphological acceleration technique included in the model, 2) the difference between the results of 2DH and 3D models, 3) the effect of including a schematised wave climate in the morphological computation, and 4) the sensitivity of the model to selecting two alternative bed roughness formulations. Finally, we present the results of two simulations of theoretical situations previously discussed in the literature. In the first theoretical test the new model is compared with an existing 2DH model for the case of a propagating Gaussian hump, which deforms into the well-known star-shape discussed by de Vriend (1987). The test series is then concluded with the offshore breakwater case discussed by Nicholson et al. (1997).

The tests performed to validate the sediment version of Delft3D-FLOW can be divided into three categories: 1) Simulations for which an analytical solution exists; 2) simulations of physical experiments and prototype situations that have reliable initial, boundary, and final conditions; and 3) simulations of situations where the results can only be compared with theoretical considerations and results produced by other computer models. Overall validation of the model in a complex and dynamic estuarine environment is described in Chapter 4.

3.2   Comparison with Analytical Solutions

3.2.1   Equilibrium conditions

Under equilibrium (i.e. stationary and uniform) conditions the advection diffusion equation reduces to

$$cw_s + D_V \frac{dc}{dz} = 0 \tag{3.1}$$

Under the assumptions of a constant sediment fall velocity and a parabolic sediment mixing profile this equation has the solution

$$\frac{c}{c_a} = \left[\frac{a\,(h-z)}{z\,(h-a)} \right]^\lambda \tag{3.2}$$

where the suspension number, $\lambda = \frac{w_s}{\beta \kappa u_*}$. This is commonly referred to as the 'Rouse' sediment concentration profile.

A series of tests were performed simulating the suspended sediment transport in a very long (8 km) straight flume with a constant water depth of 5.0 m. The depth-averaged velocity in the flume was 2.0 m/s. Figure 3.1 shows the sediment concentrations calculated 6 km from the upstream boundary where clear water enters the flume. The results of 5 simulations performed using differing numbers of (logarithmically spaced) layers, and different turbulence closure models (TCMs) are presented, superimposed on the analytical Rouse profile. It is clear that, under equilibrium conditions, the computed sediment concentrations are not sensitive to either the number of layers or the chosen TCM and that all computed results lie close to the analytical Rouse profile.

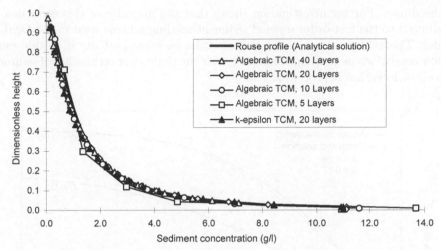

Figure 3.1 *– Equilibrium sediment concentration profiles computed 6 km down a long flume showing the effect of varying layer spacing and choice of turbulence closure model (TCM).*

3.2.2 Suspended sediment transport development

Another interesting case, for which an analytical approximation exists, is the development of suspended sediment transport at the upstream end of a channel with an initially clear flow. In this case, the flow pattern is stationary, however, the suspended sediment transport rate increases with distance down the channel until equilibrium conditions are achieved. This situation was simulated using the Delft3D-FLOW module and the results compared with the analytical solution of Hjelmfelt and Lenau (1970). The simulation parameters were: $h = 1.0$ m, $\overline{U} = 1.5$ m/s, $C = 47$ m$^{0.5}$/s, $a = 0.05$ m.

The simulated flume consisted of 120 one-metre grid cells in the longitudinal direction and 50 logarithmically spaced layers in the vertical. The simulations were performed using the algebraic turbulence closure model. The density effects of the suspended sediment were neglected, and a computational time step of 1.5s was used. Sediment transport, but not morphological change, was allowed to occur during the simulation. Three simulations were performed with median sediment diameters of 184, 130.5, and 67.5 μm respectively. These sediment sizes were chosen to produce dimensionless suspension number, λ, values of 0.5, 0.3, and 0.1 for the three simulations. This allowed direct comparison with the analytical results published by Hjelmfelt and Lenau.

Figure 3.2 is a vertical longitudinal slice through the first simulation ($\lambda = 0.5$) and shows contours of equal sediment concentration. In this plot all variables are non-dimensionalised in order to compare with the analytical results. The dimensionless variables used in the plot are defined as follows: $\lambda = \frac{w_s}{\beta \kappa u_*}$, $X = \frac{\beta \kappa u_* x}{\overline{U} h}$, $Z = \frac{z}{h}$, $C = \frac{c}{c_a}$, and $A = \frac{a}{h}$.

The figure shows that the simulation results reproduce the analytical solution well, although suspended sediment concentrations are over-estimated slightly (less than 10%) as equilibrium conditions are approached toward the downstream end of the flume. Further investigation shows that the majority of this error can be attributed to the first-order upwind sediment settling scheme used in the Delft3D model. This scheme is retained, however, as its excellent stability in a wide range of flow conditions more than compensates for the slight over estimation of sediment concentrations that it produces.

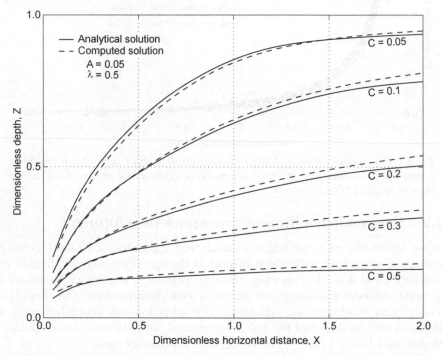

Figure 3.2 – Contours of equal sediment concentration along a flume showing the adaptation of the suspended sediment concentration from an initially clear flow (after Hjelmfelt and Lenau, 1970).

Figure 3.3 illustrates the gradual development of the sediment concentration profile along the length of one Delft3D model simulation. It can be seen that the profile develops smoothly and that equilibrium conditions have still not been reached by a distance of $x/h = 100$.

Figure 3.4 shows a longitudinal profile of the depth-averaged suspended sediment concentration computed in three simulations using different sediment grain sizes. Again the computed results compare well with the analytical solution of Hjelmfelt and Lenau. The simulations were repeated using just six σ-layers in the vertical and the results were very similar (maximum error of 16% compared to the analytical solution).

3.2.3   Equilibrium slope of a straight flume

In this experiment a relatively short straight flume with a moveable bed is simulated. At the upstream boundary a constant-discharge boundary condition is applied and the flow enters the flume carrying the local equilibrium suspended

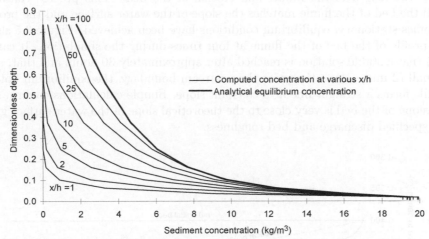

Figure 3.3 *– Suspended sediment concentration profiles at various distances along a flume showing the progressive development of an equilibrium sediment concentration profile.*

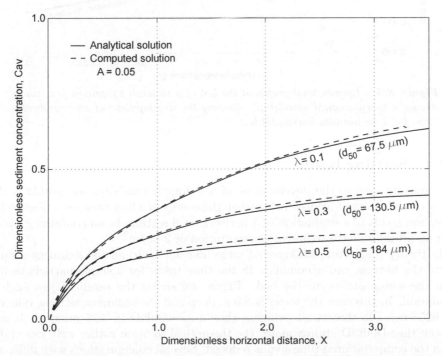

Figure 3.4 *– Development of the depth-averaged suspended sediment concentration along a flume for three different sediment sizes (after Hjelmfelt and Lenau, 1970).*

sediment concentration profile. At the downstream end of the flume a constant water level is specified. As the bed of the flume is initially horizontal, an accelerating flow is created. This in turn causes an increasing sediment transport rate along the length of the flume, and erosion of the bed. This process continues until the bed of the flume matches the slope of the water surface and the process becomes stationary; equilibrium conditions have been achieved. Figure 3.5 shows the profile of the bed of the flume at four times during the simulation. It can be seen that a stable solution is reached after approximately 30 hours and that, after a small (2 mm) adaptation near the upstream boundary, the equilibrium bottom profile forms a straight line at a constant slope. Simple calculations confirm that the slope of the bed is very close to the theoretical slope of the water surface given the specified discharge and bed roughness.

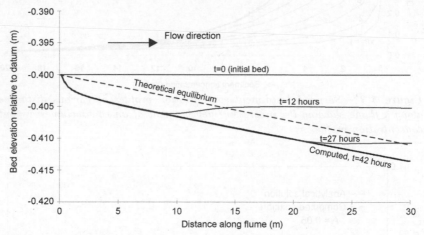

Figure 3.5 – *Longitudinal profile of the bed of a simulated flume at four times during a morphological simulation, showing the development of an equilibrium slope from an initially horizontal bed.*

3.2.4   Settling basin

The opposite case to the development of a sediment concentration profile is the falling of sediment out of suspension when the energy of a flow decreases. Figure 3.6 shows the result of a simulation of a perfectly still settling basin containing water with an initial uniform sediment concentration of 2 kg/m^3.

In theory (neglecting background molecular diffusion) all the sediment should fall to the bottom and accumulate in the time taken for a single particle to fall from the water surface to the bed. Figure 3.6 shows the result of just such a simulation. In this case the water is 5.0 m deep and the sediment settling velocity is 0.0257 m/s. In theory, all sediment should accumulate in 3.25 minutes. It can be seen that Delft3D approximates the theoretical solution rather well, especially when the computational time step is reduced. Several configurations with different layer spacings were tested and little sensitivity was discovered. In all tests, the total quantity of sediment available in the flow accumulates at the bed; continuity

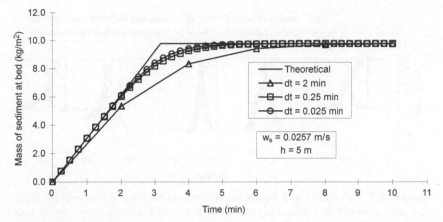

Figure 3.6 – *Results of the settling basin test showing accumulation of initially suspended sediment at the bed. Three simulations show the influence of the computational time step on the computed result.*

of sediment is preserved.

3.3 Comparison with Physical Measurements

3.3.1 Trench migration experiment

In this experiment, water flows across a steep-sided trench cut in the sand bed of a flume. The water reaches the upstream edge of the trench carrying the equilibrium suspended sediment concentration profile. As the flow decelerates over the deeper trench some sediment is deposited. Sediment is then picked up by the accelerating flow at the downstream edge of the trench. Due to the spatial difference between the areas of deposition and erosion, the trench appears to migrate downstream. Figure 3.7 shows the initial situation before the trench starts to deform. Both the results of measurements carried out by van Rijn (1987) and the computed results of Delft3D-FLOW are presented. The significant changes in both flow velocity and sediment concentration profiles, measured as the flow crosses the trench, are well represented in the Delft3D simulation.

Figure 3.8 shows the measured and computed position of the trench after 15 hours. The computed simulation was performed using a hydrodynamic simulation which ran for just 5 minutes of active morphology, after an initial spin-up period. A morphological acceleration factor (morfac) of 180 was applied to achieve the required 15 hours of morphological development. Sensitivity tests using morfacs of 90 and 45, with correspondingly lengthened hydrodynamic simulations, were also performed. The results were insignificantly different from those presented here. It can be seen that the trench has been reduced to approximately one half of its initial depth, and has migrated about 3 m downstream. The computed result is in very good agreement with the measurements. All computational tuning parameters were left at their default values.

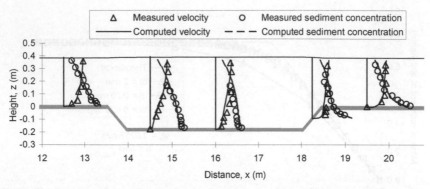

Figure 3.7 – *Measured and computed velocity and sediment concentration profiles over a trench in a flume before significant morphological development takes place (after van Rijn, 1987).*

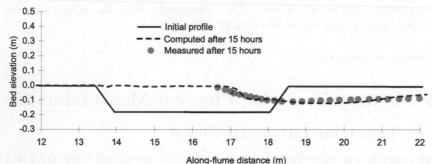

Figure 3.8 – *Measured and computed trench profile after 15 hours (after van Rijn, 1987).*

3.3.2   Curved flume experiment

In this experiment another flume test was simulated. In this case the flume begins with a straight inflow section 7 m long followed by a bend of 140° with a radius of curvature of 12 m and a straight outflow section 11 m long. The width of the flume is 1.5 m. The flume has a mobile sand bed and a sand pump was used to re-circulate the sand deposited at the downstream end of the flume. The flume was run at a constant discharge for approximately two weeks until an equilibrium state was reached. Over a period of three days 25 sets of bed-level readings were taken at 10 points across each of 45 cross-sections. The number of repeated readings was judged to be sufficient to remove the effects of slowly propagating bed forms and to provide a reliable estimate of the time-averaged bed level at each point (Figure 3.9).

The important parameters governing the experiment were as follows: $Q = 0.047$ m^3/s, $\bar{h} = 0.08$ m, $\bar{u} = 0.39$ m/s, $I = 2.36 \times 10^{-3}$, $C = 28.4$ m$^{0.5}$/s, and $D50 = 450$ μm where I is the average longitudinal slope of both the water surface and bed. For a detailed description of the model features and operation, reference is made to Struiksma (1983).

The left panel of Figure 3.9 shows the equilibrium water depths measured in the

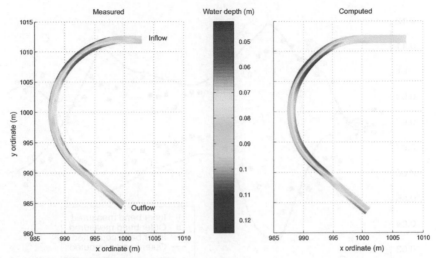

Figure 3.9 *– Measured and computed equilibrium water depths in a curved flume (after Struiksma, 1983).*

experiment. The measurements clearly show the effect of the 3D spiral flow in the bend on the bathymetry, as well as a less distinct oscillation caused by the entire depth-averaged flow 'bouncing' from one wall of the flume to the other, around the bend, and continuing downstream. The experiment was simulated using the online sediment version of Delft3D-FLOW using a curvilinear grid consisting of 10 x 93 grid cells and ten σ-layers in the vertical. The numerical model was run using the geometry and parameters described above, using bed-load sediment transport only, until the bathymetry became reasonably stable (33 morphological hours). The results can be seen in the right panel of Figure 3.9 and in Figure 3.10.

Again, the computed result is pleasingly close to the measured bathymetry. We emphasise that this result was achieved with all sediment and flow parameters set at default values with the three following exceptions: 1) the horizontal fluid viscosity was reduced to a constant value of 0.001 m^2/s to reflect the rather fine computational grid, 2) the transverse bed-slope effect factor was increased from 1.5 to 2.0, and 3) the bed roughness was specified as a constant roughness height of 0.025 m, rather than use the Chezy value of 28.4 m$^{0.5}$/s specified in the description of the experiment. We found that, in this case, it is essential to specify a constant roughness height as the depth changes significantly across the flume. When the Chezy roughness formulation is used the roughness height becomes a function of the water depth, thereby introducing a significant variation in roughness across the flume. In this case, specifying a Chezy roughness value appears to have a strong damping effect on the bathymetry - not in line with the experimental results. The bathymetry and long-sections for the same simulation performed using the Chezy roughness coefficient are presented in Figures 3.11 and 3.12.

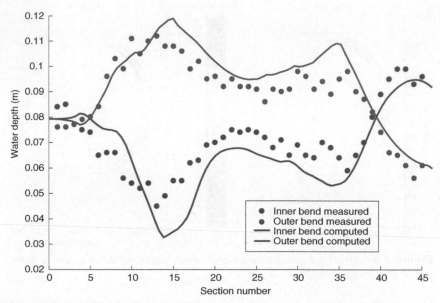

Figure 3.10 – *Longitudinal profiles of measured and computed equilibrium bed levels 0.225 m from each side wall of a curved flume.*

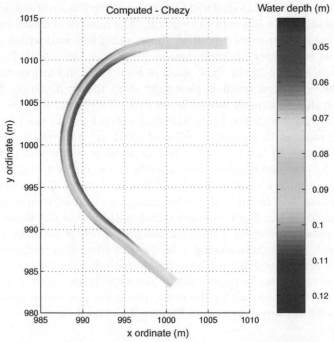

Figure 3.11 – *Equilibrium water depths in a curved flume computed using a Chezy roughness coefficient.*

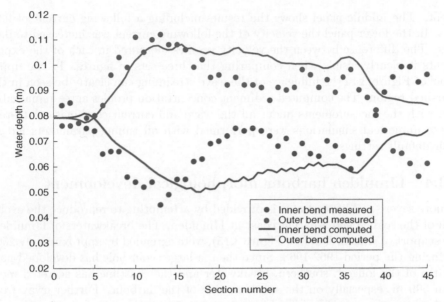

Figure 3.12 *– Longitudinal profiles of measured and computed equilibrium bed levels 0.225 m from each sidewall of a curved flume. Computation uses a Chezy roughness coefficient.*

3.3.3   Wave and current flume experiment

This experiment conducted by Dekker and Jacobs (2000) in a 45 m long wave flume investigated the velocity and sediment concentration profiles produced by three combinations of waves and currents acting over a sand bed (D50 =165 μm). All experiments were performed using random waves (H_s = 15 cm). The strength of the current, which was always in the same direction as the waves, was varied from one experiment to the next. The layout of the flume used in their experiment is reproduced in Figure 3.13; the water depths are indicated in centimetres. The sediment measurements were made 14 m downstream from the start of the mobile sand bed.

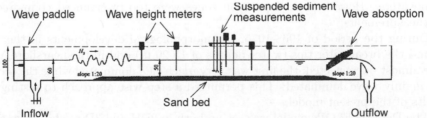

Figure 3.13 *– Arrangement of wave and current flume experiment (after Dekker and Jacobs, 2000).*

The results of three of Dekker and Jacobs' experiments and the corresponding computer simulations using the $k - \varepsilon$ TCM and 20 σ-layers are presented in Figure 3.14. The results using the algebraic TCM were very similar. Figure 3.14 upper panel shows the profiles recorded for the wave-only (no net current) experiment. The middle panel shows the results including a following current of 0.26 m/s. In the lower panel the velocity of the following current was increased to 0.40 m/s. The difference between the velocity profiles measured in each of the experiments is clearly visible when comparing the three sets of figures. In the upper panel of Figure 3.14 the influence of the wave streaming can clearly be seen in the near-bed region. The computed sediment concentration profiles agree remarkably well with the measurements under all the wave and current combinations tested. These numerical simulations were performed with all tuning parameters left at their default settings.

3.3.4   IJmuiden harbour morphological development

A more severe real-life test case is provided by attempting to reproduce the evolution of the seabed and adjacent coast at IJmuiden. The breakwaters of IJmuiden, the seaport of Amsterdam (see Figure 3.15), were extended by approximately 2500 m during the period 1962-1968. Since then, a large scour hole has developed near the tip of the longest, southern, breakwater and the coastline has accreted more than 500 m, especially on the southern side of the harbour. Further away from the harbour the coast has suffered erosion.

Regular bathymetric surveys of the area have been performed and dredging data has been collected. Roelvink et al. (1998), compiled, digitised where necessary, and tailored these data for model validation. Data have been collected over a period of 28 years, however, in this test we focus on the first 8 years of morphological development from 1968 to 1976. The shore-parallel tidal motion in the area is well documented and several operational models exist which can be used to generate boundary conditions for a detailed model of the vicinity of the harbour. Directional wave data are available from nearby stations in approximately 20 m of water depth.

Roelvink et al. (1998) compared several model simulations. These simulations used the 2DH (offline) mode of the Delft3D model system (Delft2D-MOR) and transport formulations by Bijker (1971) and were compared with a simplified transport update scheme. Qualitatively, a reasonable agreement was found for the development of the scour hole and deposition zones north and south of the breakwaters. However, the deposition zones were too pronounced compared to the measurements.

During the period of 1968-1976, the morphological developments in the area around the breakwater tips (water depths of 15-20 m) and in the nearshore areas were almost unconnected; the first was very much tide dominated while the second was mainly wave dominated. This permitted a step-wise approach to testing the results of the present model.

The Delft3D-FLOW model was set up both in 2DH and 3D mode. First, 2DH simulations were performed to test the effect of the morphological factor. Then the model was extended to 3D and simulations were performed both with and without waves. Finally, a simulation with a different bed roughness formulation was made to check the sensitivity of the model to this important parameter. Table 1 gives an overview of the simulations performed.

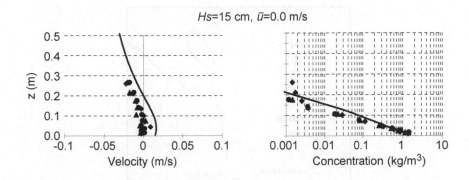

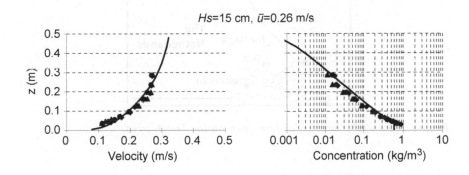

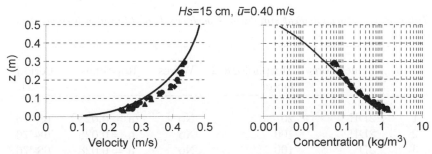

Figure 3.14 *– Top panel: Measured and computed velocity and sediment concentration profiles for waves with no net current. Middle panel: Measured and computed velocity and sediment concentration profiles for waves with a moderate following current. Lower panel: Measured and computed velocity and sediment concentration profiles for waves with a stronger following current.*

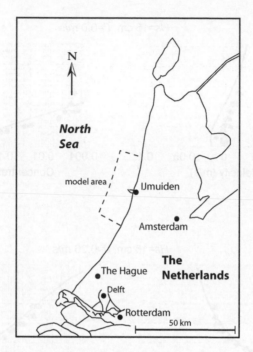

Figure 3.15 – *The location of IJmuiden on the Dutch coast.*

Table 3.1 – *Morphological simulations performed for IJmuiden breakwater extension*

Sim. no.	Dimensions	Morphological factor	Waves	Roughness	Period
1	2DH	20	No	n = 0.028	'68-'76
2	2DH	100	No	n = 0.028	'68-'76
3	3D	100	No	n = 0.028	'68-'76
4	3D	100	Yes	n = 0.028	'68-'71
5	3D	100	Yes	C_d = 0.0027	'68-'71

A curvilinear grid was constructed with a good (approximately 100m) resolution near the harbour entrance and near the coast, but gradually coarsening towards the model boundaries, which were located some 10-15 km from the harbour. A single representative tide was selected based on initial transport computations over a spring-neap cycle. The selected tide had an amplitude of 1.1 times the average tidal amplitude.

All simulations were made with a time step of 2 minutes. In order to simulate 8 years of morphological change the simulation with a morphological factor of 20 required the flow computation last for 280 tidal cycles. With a morphological factor of 100 only 56 tidal cycles had to be simulated. In the 3D case, 8 non-equidistant layers were chosen and the algebraic TCM was applied.

In Figure 3.16 the morphological changes computed in the first three simulations are compared with the measured sedimentation and erosion. A comparison between the simulations with different morphological factors shows no discernable difference. A comparison of these two simulations is presented in more detail in Figure 3.17 which shows the time-evolution of the bed elevation for two points, one located in the scour hole and the other in the southern deposition lobe. It is clear that the morphological changes within a tidal cycle, even scaled up by a factor of 100, are still small compared to the water depth and the longer-term morphological trend.

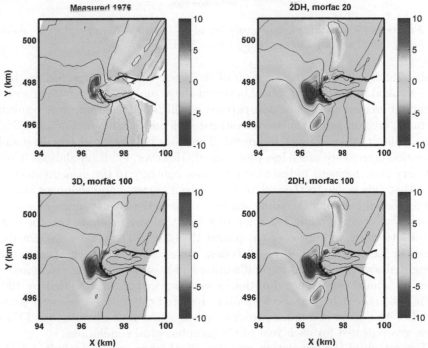

Figure 3.16 – *IJmuiden, measured and computed sedimentation (red) and erosion (blue) patterns for 2DH and 3D morphological simulations of the period 1968-1976 (no waves).*

The 2DH simulation results are very similar to those presented by Roelvink

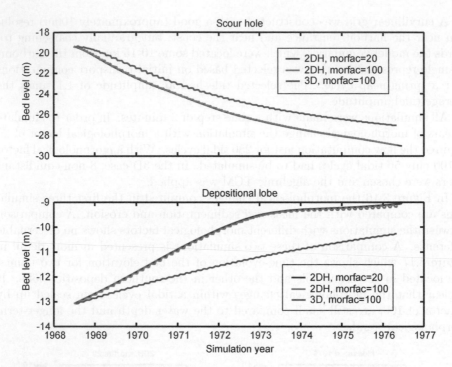

Figure 3.17 – *IJmuiden, time evolution of bed elevation for a point in the scour hole (upper) and a point in the deposition area (lower).*

et al. (1998), although the location of the scour hole has improved as the thin dams applied in the present model better fit the actual breakwater alignment. The computed erosion and deposition pattern is still too strong however, resulting in too much erosion and strongly overestimated deposition lobes. For the 3D simulation, this situation is much improved. The scour hole becomes less deep and the depositional areas are much less pronounced. However, in all simulations it is clear that very little happens in the nearshore area, contrary to the measurements. It is likely that this is due to the lack of wave-driven currents and sediment transport.

For the simulations including waves, a schematised time series of wave conditions was created. The basis for this was the wave climate recorded at the Euro Platform wave buoy over the period 1979-2001. The total wave climate was binned in 2-metre H_s, 30-degree direction, wave classes. For each of these classes, a weighted average H_s and mean direction and wave period were computed. The expected annual occurrence duration of each wave class was divided by 100 (reflecting the morphological acceleration factor) and the wave classes were arranged one after the other in random order to make a one-year time series. This time series was repeated for each year of the morphological simulations.

The set-up of the simulation was simple: the flow model including sediment transport and morphological changes was run for one hour (100 hours of morphological change), then the waves were updated using the updated bathymetry and water levels from the flow model, after which the flow model ran with updated waves. This cycle was repeated for 56 tidal cycles. The SWAN model was used to compute the waves on a curvilinear grid identical to the flow grid except for extensions at the southern and northern ends. As these simulations are somewhat more demanding they were carried out for only 3 years. The morphological evolution was smooth however and longer simulations are possible.

In Figure 3.18 the results are compared with the observed erosion and sedimentation over the period 1968-1971. Simulation no. 4 (lower-left panel) had the same roughness settings as the other simulations, viz. a Manning coefficient of 0.028, which was adopted from the regional model this model was nested in. Clearly, much more is happening in the nearshore than in the simulations without waves, but it still rather underestimates the observed accretion near the breakwaters. The Manning coefficient used is fairly high, but leads to an especially strong increase in bed friction in the shallow water near the coast. This leads to a sharp reduction in longshore velocity and sediment transport. In the last simulation, a constant C_d value of 0.0027 was applied (following the findings of Ruessink et al., 2001). This setting gives similar roughness in deep water but much less in shallow water. The results of this simulation (lower-right panel) show a marked increase in transport near the coast and much stronger deposition beside the southern breakwater, in line with the observations. The change in roughness coefficient has, however, also had a marked impact on the depth of the predicted erosion hole, which is now rather shallower than observed. This observed sensitivity of the predicted morphology to selected bed roughness coefficient clearly indicates an area requiring further research in order to develop a reliable predictor of bed roughness under the combined influence of waves and current.

3.4   Comparison with Other Numerical Models

3.4.1   Hump test

This test checks a theoretical test problem previously discussed by de Vriend (1987). An east-west oriented rectangular channel 10 m deep, 10 km wide, and 20 km long was subjected to eastward flow with a velocity of 1 m/s. The bed of the channel consists of sand (d50 = 200 μm) and contains a Gaussian hump with a radius of 1 km and initial height of 5 m (see the top two panels of Figure 3.19).

In the centre-left panel of Figure 3.19 we see the evolution of the hump according to a standard 2DH Delft2D-MOR simulation, using van Rijn bed-load and suspended-load transport for a duration of 200 days. Typical for this type of evolution is that the top of the hump moves in the direction of the flow, but two lobes extend obliquely at the same time. On either side of the hump, and in front of it, there are regions of erosion.

The remaining panels of Figure 3.19 present the results of three 3D simulations using the online morphology version of Delft3D-FLOW. It is clear that the 3D results are qualitatively similar to those of the 2DH simulation. However, the development of the lobes is less pronounced in the 3D simulations and the bathymetry predicted after 200 days is somewhat smoother in the 3D case. Comparison of the bed-shear stress distributions for the 2DH and 3D simulations at

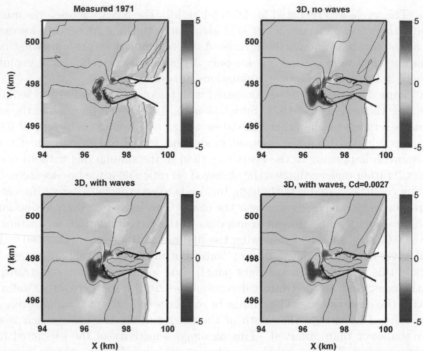

Figure 3.18 *– IJmuiden, measured and computed sedimentation (red) and erosion (blue) patterns for 3D morphological simulations of the period 1968-1971.*

the start of the morphological changes shows that the gradients in bed-shear stress are significantly reduced in the 3D simulation due to deformation of the logarithmic velocity profile. We expect that this smoother distribution of bed-shear stress causes the smoother development of the bathymetry in the 3D simulation.

Comparison of the results of the 3D simulations also serves as another useful test of the morphological acceleration factor included in Delft3D-FLOW. Over these simulations the morphological acceleration factor increases by a factor of 25 (thereby decreasing the required hydrodynamic simulation duration by the same factor). Little difference can be seen in the resulting bathymetry after 200 morphological days. Therefore it is clear that the use of even a rather high morphological acceleration factor has little impact on the development of the morphology in this situation. It is important to stress, however, that appropriate morphological acceleration factors must be chosen and tested on a case-by-case basis.

3.4.2   Offshore breakwater case

A more complex test of the combined modelling of waves, currents and morphological changes in 3D is provided by the offshore breakwater test case reported in Nicholson et al. (1997). Nicholson et al., used this test case to compare five different 2DH morphodynamic models, including the standard version of DELFT2D-MOR.

The test geometry consists of a long straight coastline with a planar sloping

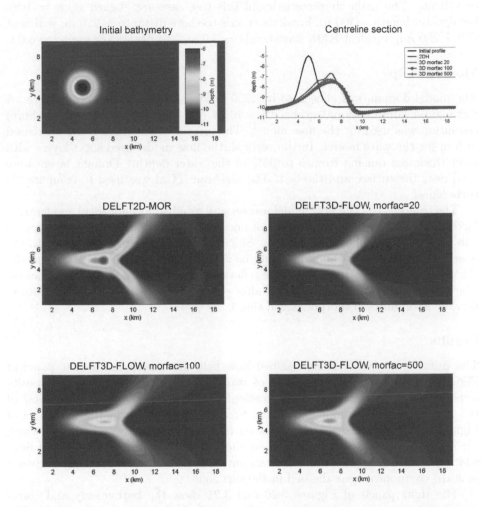

Figure 3.19 *– A rectangular channel containing a Gaussian hump is subjected to an east to west flow for 200 days. Top-left panel shows initial bathymetry; top-right panel shows centerline sections through the hump initially; and after each simulation. The remaining panels show aerial views of the deformed bathymetry predicted by four different model simulations.*

beach. Offshore from the beach lies a shore-parallel surface-piercing impermeable breakwater. The only driving force in the test is provided by the incoming waves, which enter perpendicular to the coast. Longshore gradients in wave set-up drive a double circulation pattern in the nearshore flow that tends to bring sand into the sheltered area behind the breakwater. This leads to the formation of a tombolo or salient. The main characteristics of this test case are: beach slope = 1:50, breakwater length = 300 m, breakwater axis-to-shore distance = 220 m, sediment d50 = 250 μm, incident RMS wave height = 2.0 m, and peak wave period = 8.0 s.

Model set-up

The model domain was chosen to be 1300 m longshore by 700 m cross-shore. A rectangular computational grid with a 20 m (longshore) by 10 m (cross-shore) resolution was used for the flow model. The cross-shore resolution was increased to 5 m for the wave model. In the vertical, the flow model used six σ-layers with layer thickness ranging from 5 to 35% of the water depth. Thinner layers were used near the surface and the bed. The algebraic TCM was used to compute 3D turbulence.

The time step for the flow model was set at 6 s and a morphological acceleration factor of 24 was used. This meant that one hour of flow computation represented one full day of morphological change. Since the waves were perpendicular to the beach, all flow model boundaries could be closed boundaries. For the wave model the side boundaries were chosen to be reflecting in order to minimise disturbances. The wave computation was updated after every ten flow time steps (1 minute in flow, or 24 minutes of morphological time).

Results

The initial bathymetry and near-bed flow field are shown in the left panel of Figure 3.20. The expected pattern of two circulation cells rotating in opposite directions, driven by wave set-up gradients is clearly visible. A closer view of the initial situation at one end of the breakwater is shown in the left panel of Figure 3.21. This figure also shows the current vectors at the water surface, which are clearly different from the flow near the bed. This is due to: 1) helical flow, which pushes the upper-layer velocities outward; and 2) undertow, which gives a seaward component near the bed in the surf zone.

The right panels of Figures 3.20 and 3.21 show the bathymetry and corresponding flow patterns after 72 hours of morphological change have occurred. The model is clearly demonstrating a tendency to accumulate sediment behind the breakwater, with the salient reaching from the shoreline almost out to the breakwater. The crest of the salient has reached a level of one metre below the still water level, which is likely to be as high as it can grow in the absence of a tidal range. Significant erosion occurs near the beach on either side of the breakwater and a deep scour channel is formed inshore of the tips of the breakwater.

At first sight, the results are quite similar to the results of the 2DH models, given in Nicholson et al. (1997). A difference however is that the accretion behind the breakwater occurs at the expense of stronger erosion of the beach at the edges

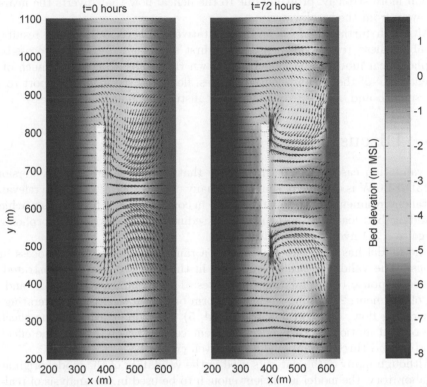

Figure 3.20 – *Overall view of initial (left panel) and final (right panel) bathymetry and near-bed flow fields computed for an offshore breakwater subject to shore-perpendicular waves.*

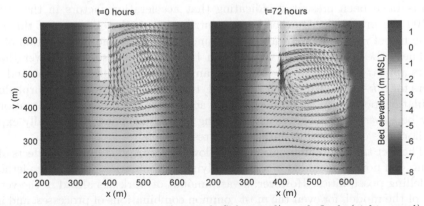

Figure 3.21 – *Detail view of initial (left panel) and final (right panel) bathymetry, near-bed flow field (black arrows) and near surface flow field (red arrows).*

of the lee zone, due to undertow. Also, the scour holes at the breakwater tips develop more strongly, probably due to the helical flow that diverts the near-bed flow away from the breakwater.

Although further analysis and quantitative comparison of model results are desirable, these results are a promising first step for a fully 3D process-based morphological model. The model has shown the ability to combine much of our present state-of-the-art knowledge of waves, flow, and sediment transport to produce smooth and, at least, qualitatively realistic morphological changes.

3.5   Discussion

The validation cases presented here show that the online morphology version of Delft3D-FLOW is capable of simulating many of the processes that are relevant in coastal environments, both separately and in combination. This has been achieved by adding bed-load and suspended-load sediment transport and morphological change to a 3D hydrodynamic flow model.

The model has been validated across a range of processes and process interactions. The validation studies reported in this chapter have demonstrated the model's response to the following processes: 1) entrainment, transport, and settling of sediment, 2) varying levels of uniform bed shear stress, 3) accelerating and decelerating flow, 4) spiral flow in a bend, 5) bed slope effects, 6) the effects of wave orbital motion on suspended sediment concentration, and 7) the effects of undertow and three-dimensional wave-driven currents.

Although many of the datasets used in the validation of the model originate in the laboratory, the model is efficient enough to be used in the analysis of real-life, prototype-scale, situations. The suspended transport and bed-load transport are computed using the same time step as the flow model, but morphological changes can be accelerated by use of a morphological acceleration factor. It has been demonstrated that for simple cases very high morphological factors can be used without significantly changing the solution. For tidal and wave-driven situations results have been presented indicating that acceleration factors in the order of 50-100 are practical. Another factor contributing to the efficiency of this model is the use of σ-layers. With a sensible (smooth, logarithmic) distribution of layer thickness the concentration vertical can be efficiently resolved. Fewer than 10 layers appear to be required when using an algebraic turbulence model, and 10-20 layers for a $k - \varepsilon$ model. When combined, these factors allow useful morphological simulation periods to be covered within acceptable runtimes. This situation can only be expected to improve as increasing computing power continually expands the horizons of simulation duration and resolution.

The fact that any combination of 3D flow processes, including the effects of sediment, can now be included in a morphodynamic simulation opens up a plethora of modelling possibilities. However, a vast amount of work lies ahead in the verification of the model, for even the most common combinations of processes, and in the refinement and extension of the model components. A clear example of the need for further development is provided by two of the tests presented in this chapter, which show a marked sensitivity to the chosen bed roughness parameter. It is well known that in reality bed roughness is highly variable both in space and in time. In this respect reality is far more complex than the formulations employed in the present model and this is an area that clearly requires both an advancement of the state-of-the-art of our understanding and the parallel development and validation of reliable model formulations.

Future research efforts will need to focus on the definition of further test cases, the generation of comprehensive datasets for validation of specific combinations of processes, the refinement and extension of model formulations, and the testing and application of the model.

Chapter 4

Medium-term Modelling of Willapa Bay

In this chapter the numerical morphological model developed in Chapter 2 and validated against test cases in Chapter 3 is applied to a real-life coastal erosion problem. The objectives of this chapter are to:

1. Develop and test methods for applying the model to modelling morphological change in a complex coastal environment over a duration of several years.

2. To validate the hydrodynamic and wave model components against detailed process measurements made specifically for this purpose.

3. To calibrate and validate a complex coastal morphological model.

4. To understand the physical processes responsible for the observed coastal erosion and make predictions of future morphological change.

The location selected for applying the model is the entrance to Willapa Bay in the state of Washington, USA. The work described in this chapter was funded by the US Geological Survey (USGS) and the work was performed based at the USGS Menlo Park, CA campus during 2002 to 2005. The support of the USGS is gratefully acknowledged.

4.1 Introduction

4.1.1 Background

Coastal residents living at North Cove and Tokeland, Washington, USA (See Figures 4.1 and 4.2) have witnessed extremely rapid beach recession associated with northward migration of the entrance channel to Willapa Bay (Figure 4.2) and destruction of Cape Shoalwater (Andrews, 1965; Terich and Levensellar, 1986; Dingler and Clifton, 1994; Kaminsky et al., 1999). The rapid retreat of Cape

Shoalwater since the late 1800s destroyed lighthouses and continues to demolish homes (Figure 4.3), threatens other coastal properties, and reduces the tribal lands and shellfish resources at the Shoalwater Bay Tribe Reservation. The recent breaching of Empire Spit (also known as Graveyard Spit) and loss of barrier protection from storm-surge flooding and direct attack from storm waves in the Pacific remains a constant concern of the Tribe, local residents, and property owners.

This modelling study formed part of a larger (unpublished) study conducted by the USGS and the Washington Department of Ecology to provide technical information to assist the US Army Corps of Engineers District Office assess the feasibility of providing erosion protection. Information gathered by the study was also used to inform tribal members and residents of Tokeland, as well as local, state, and federal government resource managers. Specific objectives of the broader USGS study were to:

1. Quantify the historical trends of shoreline erosion and channel migration along the northern shore of Willapa Bay;

2. Better understand the processes responsible for shoreline erosion and channel migration, and;

3. Predict the future position of the shoreline and channel location.

These objectives are addressed through a series of technical studies. The technical studies were designed to provide answers to several questions:

1. Why was the channel migrating so rapidly in the past?

2. What processes (tide, waves, etc.) were responsible for the shoreline erosion and channel migration?

3. Why has the erosion slowed or stopped? and

4. What will likely happen in the future?

The process of developing answers to these questions was intended to provide insight into the dynamic nature of Willapa Bay and provide important information to guide the implementation of sustainable solutions.

The technical studies carried out for the USGS study were:

1. Geological setting (published in Morton et al., 2002)

2. Historical shoreline and land-surface changes

3. Historical channel and entrance migration

4. Recent shoreline and channel migration

5. Measurement of oceanographic processes, including waves, currents, and sediment transport

6. Modelling of sediment transport and resulting morphological change (this study)

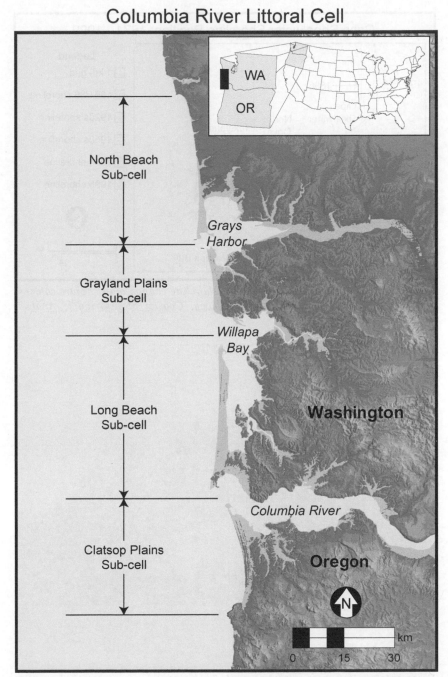

Figure 4.1 – *Location of Willapa Bay, WA, USA. Figure courtesy of Washington State Department of Ecology, Coastal Monitoring & Analysis Program*

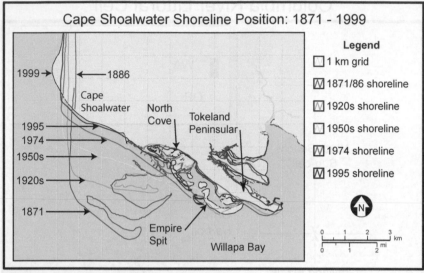

Figure 4.2 – *Cape Shoalwater historical shoreline change map. Figure courtesy of Washington State Department of Ecology, Coastal Monitoring & Analysis Program*

Figure 4.3 – *House recently destroyed due to coastal erosion at Willapa Bay. Photo courtesy of G. Kaminsky, Washington State Department of Ecology.*

The modelling study described here should be seen in this context. It was designed, in conjunction with a field measurement campaign and shoreline and bathymetry change analyses, to provide a detailed understanding of the coastal processes governing the morphological changes observed at the mouth of Willapa Bay. It was performed in order to understand the changes observed over decadal timescales and to assist making predictions of likely future large-scale morphological change of the bay entrance.

4.1.2   Related studies

Several recent studies of Willapa Bay and the adjacent coast have been conducted that are of special interest to the present work. The US Army Corps of Engineers (USACE) Coastal and Hydraulics Laboratory of the Engineer Research and Development Center published a report entitled 'Study of Navigation Channel Feasibility, Willapa Bay, Washington' that contains hydrodynamic data (waves, currents, and water levels) and modelling results (Kraus et al., 2000). The Washington Department of Ecology and the U.S. Geological Survey recently completed a regional study of the Columbia River littoral cell (Figure 4.1) as part of the Southwest Washington Coastal Erosion Study (e.g. Gelfenbaum et al., 1997; Kaminsky et al., 1999; Gelfenbaum and Kaminsky, 2000; Gelfenbaum and Kaminsky, 2002). Circulation modelling with an emphasis on the ecology of Willapa Bay has also been investigated by the University of Washington, e.g. Banas et al., 2004.

4.1.3   Coastal processes

The south-western coast of Washington is a wave-dominated, meso-tidal range region that receives sediments primarily by northward longshore transport from the Columbia River (Ballard, 1964; White, 1970; Luepke and Clifton, 1983) and the sediment in the mouth of Willapa Bay is predominantly well sorted fine sand. The high tide range (4 m) creates strong tidal currents and exchanges large volumes of water into and out of Willapa Bay. Tidal current velocities of 3 m/s have been reported at the entrance to the bay (Andrews, 1965; Clifton and Phillips, 1980; Kraus et al., 2000). At Toke Point (Figure 4.4), tide gauge measurements show that still-water elevations during winter storms can be more than 4 m above MLLW. Runup of storm waves superimposed on the high water levels would significantly increase flood elevations near the shore along Empire Spit and along exposed south-westerly-facing segments of Tokeland Peninsula.

Summer waves in the Pacific Northwest have periods of 5-10 s, whereas winter waves have periods of 10-20 s (Tillitson and Komar, 1997; Ruggiero et al., 2005). Seasonal changes in the angle of wave approach to the coast cause a reversal in longshore transport directions from predominantly southerly in the summer to predominantly northerly in the winter. Recent analyses of deep-water waves off of the west coast of the United States by Allan and Komar (2000) show that average significant wave heights range from 3.6 to 3.8 m, and the maximum significant wave heights reach 14 to 15 m. Their analyses of annual wave heights also indicate that between 1975 and 1999 the largest storm-generated waves off of the Washington coast increased in height from 8 to 12 m. The study by Allan and Komar (2000) implies that climatic changes can result in altered wave states at the coast. The combination of fine sand and high levels of tide and wave energy make for high rates of sediment transport and rapid morphological change.

4.1.4 Chapter structure

This chapter is structured as follows: Section 4.2 gives an overview of the data collected during the field measurement campaign performed to assist with model calibration. Section 4.3 describes the set-up, calibration, and validation of the numerical models used to address the questions posed. Section 4.4 presents the results of the modelling study. Section 4.5 presents conclusions regarding the past, present, and future of Willapa Bay and Section 4.6 discusses the knowledge regarding use of the new model gained from applying Delft3D-FLOW to a complex real-life erosion problem.

4.2 Field Measurement Campaign

4.2.1 Acknowledgement

Any large field measurement campaign must be a collaborative effort, and this was certainly the case for the campaign discussed here. The work of Guy Gelfenbaum, Laura Landerman, Jessica Lacy, and David Gonzales of the USGS Menlo Park, of Keith Kurrus and Kevin Redman of Evans Hamilton Inc, and of Captain Terry Larson of the fishing vessel (F/V) Tricia Rae were instrumental in performing this large-scale field campaign. Laura Landerman is also acknowledged as the original author of some of the text and several of the figures on which this section is based. Her permission to borrow heavily from her more detailed (but regrettably unpublished) description of the USGS field campaign is gratefully acknowledged.

4.2.2 Overview

As the main objectives of the modelling study were to understand sediment transport and morphological change patterns around the Willapa Bay main entrance channel and Empire Spit, the USGS measurement campaign focussed on obtaining velocity, wave, and bed and suspended sediment measurements in these locations. The locations of the USGS stations and the earlier (1998) US Army Corps of Engineers (USACE) stations can be seen in Figure 4.4.

The timing of the USGS field campaign was chosen based on an analysis of historical environmental conditions (Figure 4.5). Historically the winter months of November through January produce storms from the south, high off-shore wave heights, increased input from Willapa River, and elevated water levels. It can be seen in Figure 4.6 that these criteria were all met during the period of USGS data collection, November 2002 through January 2003.

The field campaign consisted of a combination of instrumented tripods, moorings and bottom mounts deployed to measure waves, tides, currents, density and suspended sediment concentration. Site location names were chosen based upon their location within the bay: MC - main channel, ES - Empire Spit, WR - Willapa River, NW - Nahcotta channel West, and NE - Nahcotta channel East.

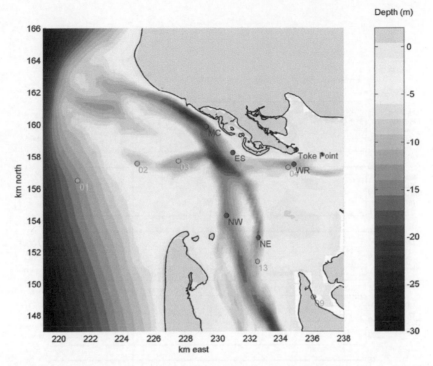

Figure 4.4 – *Map of the study area showing the USGS stations (red), USACE stations (green), NOAA Toke Point tide gauge and meteorological station (pink), and bathymetry.*

4.2.3   Coastal processes measurements

Instruments to measure pressure, temperature, salinity, current velocity profiles and velocity at near-bed points, wave height, suspended sediment concentration, and instrument orientation were deployed on 5 tripods placed on the seabed at the locations indicated in Figure 4.4. The instruments were deployed twice, and were retrieved between deployments for de-fouling, battery replacement, and to download data. The first deployment ran for four weeks from 3 November 2002 to 3 December 2002. The second deployment ran for six weeks from 4 December 2002 until 19 January 2003.

Tripods and bottom mount

The frames of the five tripods deployed during this experiment were constructed of welded aluminum tubing with an outside diameter of 7.62 cm, and were approximately 2.5 m high and 3 m wide. Figure 4.7 shows the tripod deployed at location NE. The arrangement of instruments on this tripod is typical. The lower

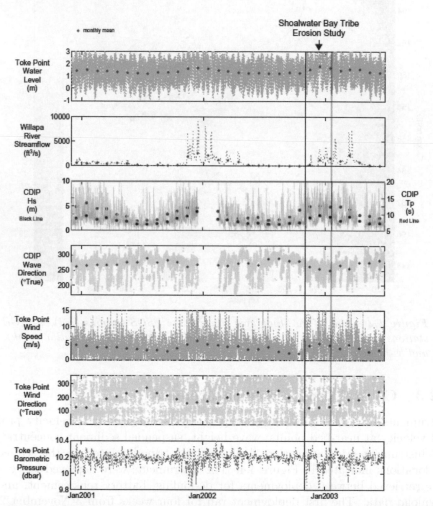

Figure 4.5 *– Time series of various environmental conditions in Willapa Bay; monthly mean is shown as a solid dot. Winter brings storms from the south, elevated water levels, increased input from the Willapa River, and large offshore waves. Figure courtesy of L. Landerman.*

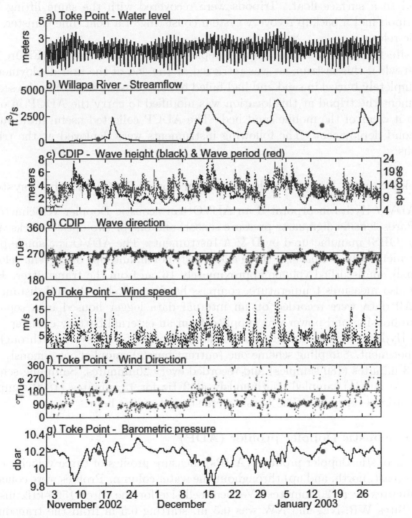

Figure 4.6 – *Time series of various environmental conditions in Willapa Bay during the three months of the USGS field campaign; monthly mean is shown as a solid dot. Figure courtesy of L. Landerman.*

crossbars were placed high on the tripods (106 cm above ground, approximately 30 cm above the ADVO and OBS sampling volumes) to minimize flow disturbance and bottom scour. The circular feet had a diameter of 30 cm, and were ballasted with removable cylindrical lead weights.

All five tripods were deployed from the F/V Tricia Rae and were lowered to the bottom using a crab block on 2.5 cm (1") polypropylene line, which was left tethered to a surface float. Tripods were recovered with the same lifting line. Each tripod had a backup recovery system consisting of a float, line canister, and acoustic release.

At site ES an ADCP was initially deployed in a bottom mount (Figure 4.8) however when this instrument was retrieved at the end of the first deployment it was completely buried in sand and had failed to collect useful data. For the second deployment the tripod at this location was modified to carry the ADCP in order to keep it clear of the mobile sand bed. The ADCP collected useful data during the second deployment. The following instruments were deployed at the tripod locations:

SonTek acoustic Doppler velocimeter ocean (ADVO) "Hydra" system

Each ADVO Hydra incorporated an ADVOcean acoustic Doppler velocimeter, a serial Paroscientific digiquartz pressure sensor, and one or two optical backscatter sensors (OBS) manufactured by D & A Instruments. The ADVO is a single-point current meter which measures three axes (east, north, and up) of current velocity in a small, undisturbed volume approximately 16 cm from the transmitter. Each ADVO also measures temperature, compass heading, pitch, roll, and distance to bed. All data were recorded on an internal data logger housed in a separate pressure housing and power was supplied from an external battery canister. The ADVO Hydra system sampling schemes were identical for all systems throughout this experiment. Sampling scheme one (current bursts) sampled continuously at 1 Hz for 3 minutes (180 samples) and recorded every 20 minutes. Sampling scheme two (wave bursts) sampled continuously at 2 Hz for 15 minutes (1800 samples) and recorded every hour. The bursts were started on the hour.

SonTek acoustic Doppler profiler (ADP)

SonTek acoustic Doppler profilers (ADPs) measure profiles of three velocity components (east, north, and up) throughout the water column. Profiles were collected for 3 minutes every 20 minutes. Vertical cell size for the three 1500-kHz instruments (Sites WR, NE, and NW was 0.5 m, starting 0.5 m from the transducer. Data were collected for 48 cells (44 at Site WR), up through the water column and past the level of the water surface. The 500-kHz instrument (site MC) had a blanking distance and cell size of 1.0 m and collected data for 30 cells. All the ADPs were set to measure as quickly as possible, and averaged profiles were recorded every 20 minutes. The instruments were set to start on the hour.

Two of the ADPs were equipped with internally mounted Druck pressure sensors and SonPro wave measurement software. Instrument failure resulted in the wave data not being collected. The ADP deployed at Site MC had no pressure

Figure 4.7 – *Photograph of the tripod deployed at Site NE during deployment 1 showing general instrument placement on the tripods.*

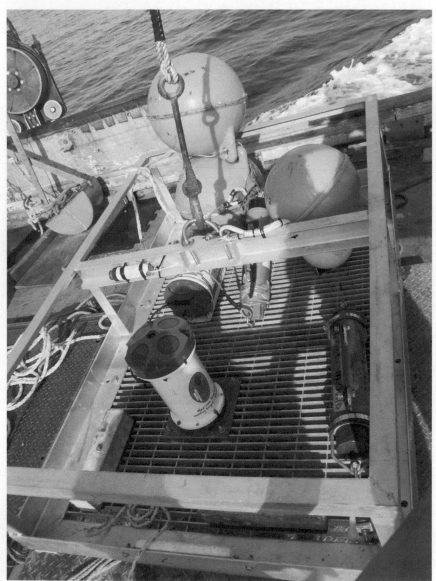

Figure 4.8 – *Photograph of the bottom mount deployed at Site ES during Deployment 1.*

sensor. The ADPs also measured and recorded water temperature, compass heading, pitch, and roll. Power was supplied from a separate battery canister. The same instruments and sampling scheme were used for both deployments.

Teldyne RDI acoustic Doppler current profiler (ADCP)

A 1200 kHz RDI ADCP Sentinel Workhorse was deployed at station ES to measure waves and current profiles. The ADCP was configured to measure and record velocities in 45 0.35 m high cells with the first cell located 0.67 m above the instrument. Current bursts consisting of 350 samples recorded at approximately 1.2 Hz were collected every 25 minutes. Wave bursts consisting of 2400 samples at 2 Hz were collected every hour. Wave data were processed into directional wave spectra and parameters by EHI using RDI's WavesMon software.

Optical backscatter sensors

Optical Backscatter Sensors (OBS) were deployed on each tripod at all sites and with some CTD casts to provide estimates of suspended sediment concentration. The OBS signal is an analogue voltage ranging from 0 to 5 V. For OBS deployed with SonTek Hydra systems the analogue signal is converted to a digital output ranging from 0 to 65535 counts by 16-bit analogue-digital converters within the individual hydra system on which the OBS were deployed. This reading of the OBS is then recorded in counts by the hydra system. The OBS responds differently depending on the type and size distribution of sediment suspended within its sampling volume, therefore a calibration for each OBS was performed with sediment collected at the station where each OBS was deployed. This calibration converts the digital value recorded by the hydra system in counts, to real-world measurements of suspended sediment concentration in g/L.

Coastal process measurement results

The field measurement campaign was extremely successful, with almost all instruments collecting useful data, the exceptions being a couple of flooded OBS instruments and the burial of the ADCP during the first deployment. The vast majority of the coastal process data are not explicitly reported here. All data were subject to QA checks before receiving further processing for comparison with the numerical models.

Typical time series data obtained from the ADVO instrument at site ES (Empire Spit) for the duration of the two USGS instrument deployments are shown in Figures 4.9 and 4.10. Data from other sites are not presented in this chapter however the results of analysed data compared to model results are presented in Section 4.3.5.

4.2.4   Bottom sediment samples

Grab samples of bottom sediments were obtained from each of the tripod deployment stations using a van Veen sampler. The grain size distribution of each

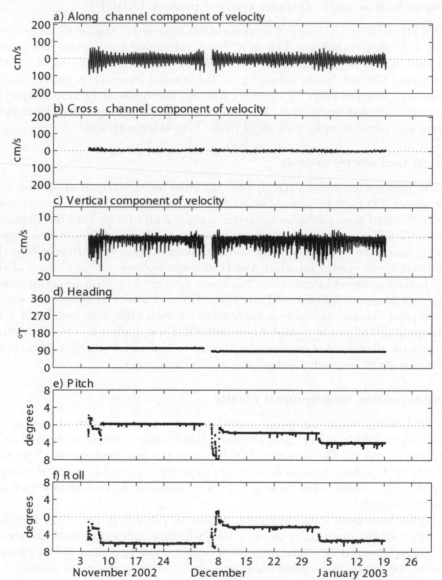

Figure 4.9 – *ADVO data from Site ES - Velocity components and tripod heading, pitch, and roll. Figure courtesy of L. Landerman.*

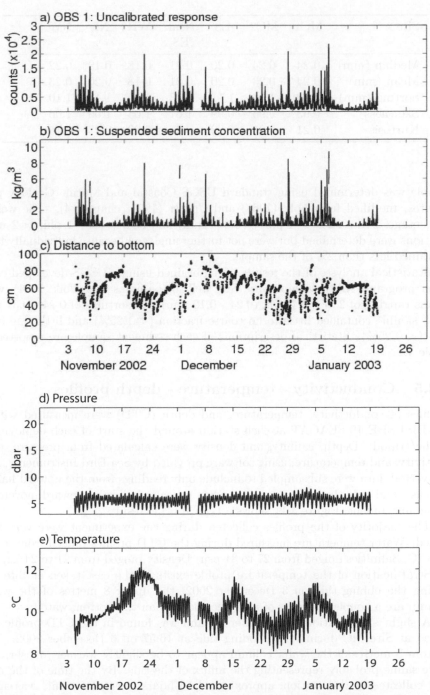

Figure 4.10 – *OBS and ADVO data from Site ES - Suspended sediment concentration from OBS and distance to bottom, pressure, and temperature measured by ADVO. Figure courtesy of L. Landerman.*

Table 4.1 – *Results of bottom sediment grain-size statistical analysis.*

Site	MC1	MC2	ES	Near ES	WR	NE	NW
Median (mm)	0.24	0.24	0.20	0.21	0.18	0.19	0.22
Mean (mm)	0.24	0.23	0.20	0.21	0.18	0.20	0.23
Sorting (mm)	1.10	1.10	1.11	1.13	1.15	1.12	1.10
Skewness	0.93	0.95	0.99	1.01	1.03	1.00	1.06
Kurtosis	0.21	0.21	0.29	0.25	0.22	0.23	0.18

sample was determined using standard USGS Coastal and Marine Geology procedures, modified from Folk (1968) and Carver (1971; chapter 4). The weight percentage of the fine (> 4 phi, < 0.063 mm) and coarse (< -1 phi, > 2 mm) fractions were determined but were not further analyzed because individually they contained less than 2% of the sample.

Statistical analyses of the results were obtained using a USGS-developed computer program. All the samples are very well sorted fine sands (Folk, 1974), with means ranging of 2.07 - 2.49 phi (0.24 - 0.19 mm) and sortings of 0.24 - 0.31 phi. Each sample contained little to no coarse fraction ($< 1.25\%$) and little fine fraction ($< 0.25\%$). Statistical descriptions of each sediment sample are reported in Table 4.1.

4.2.5   Conductivity - temperature - depth profiles

Profiles of conductivity, temperature, and depth (CTD) were measured with a Sea-Bird SBE 19 SEACAT at each station around the start of each deployment of the tripod. Depth, salinity, and density were calculated from pressure, conductivity, and temperature, using software provided by Sea-Bird Instruments, Inc. Converted data were subsampled to include only readings from the upward half of the cast, and only those data acquired when pressure indicated upward movement of the instrument. Figure 4.11 shows a typical CTD result.

The majority of the profiles collected during the experiment were very well mixed. Water temperature measured during the CTD profiles ranged from 9.4 to 10.5 °C. Salinities ranged from 27 to 31 psu. Density ranged from 21 to 24 kg/m^3.

Stratification of the temperature profile exists in the cast taken at Site NE during the ebbing tide on 3 December 2002; the upper 8 metres of the water column are a couple of tenths of a degree cooler than the bottom waters.

A slight stratification of the water column was found in the CTD profile collected at Site MC during the flooding tide at 16:57 on 6 December 2002. The bottom 3 metres of the water column appear to be slightly warmer, denser, and more saline probably representing the influx of the tide. By the time of the next cast collected at this station, approximately 3 hours later, the profile was again well mixed with the properties of the bottom waters of the previous profile.

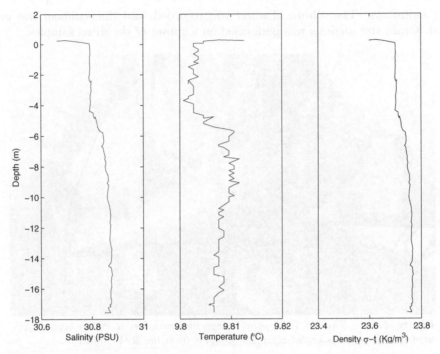

Figure 4.11 – CTD cast taken at Site NW on ebb tide. (15:00 on 5 Nov. 2002).

4.2.6   Suspended sediment sampling

Suspended sediment samples were collected with a US P-63 sampler from aboard the research vessel (R/V) Reflux. The P-63 is a 200-lb. (90 Kg) point-integrating sampler which is made of bronze and shaped like a fish which allows the sampler to point into the flow (Figure 4.12). The sampler holds one-litre sample bottles which can easily be changed. The sampler was lowered to the desired depth using a shipboard winch through the A-frame located on the stern of the boat. At the desired depth, an electrical signal is sent to the solenoid at the nozzle, which keeps the sampler watertight, telling it to open. Upon opening the solenoid water flows into the nozzle. A series of air filled chambers allow flow into the 'fish' at the rate of flow of the water. The length of time the sampler is kept open depends upon the speed of flow of the water; it is important to not to allow the bottle to overflow. Another electrical signal from shipboard tells the solenoid to close. More information about the sampler can be found in Edwards and Glysson (1999). Heights above bottom are nominal and approximate. While the sampler is meant to be point-integrating and not depth-averaging the R/V Reflux, and thereby the sampler, were moving up and down with the waves which varied in height over data collection times and sites. Additionally, the distances to bottom listed in the table are to the bottom of the sampler. The intake nozzle is actually 17 cm above the bottom of the fish, assuming the fish is oriented horizontally. The mass concentration of the samples was found by sieving the samples at 4 phi (0.063 mm); visual inspection determined the samples contained negligible amounts of finer sediments. The volume of water was recorded, and the sediment was oven dried. Grain size analysis was performed on a subset of the dried samples.

Figure 4.12 – *The US P-63 point integrating sampler, a 200-lb bronze 'fish' used to collected suspended sediment samples from the R/V Reflux.*

The mass concentration and grain size of the suspended sediment samples are plotted in Figure 4.13 with four repeatability test samples shown in magenta. It is important to remember that the heights above bottom are approximate, due particularly to choppy seas while collecting the samples; the distance of the sampler from the bottom may have varied more than a meter. Regardless, a trend can be seen in the data with suspended sediment concentrations decreasing with distance from the bottom. Anomalously high suspended sediment samples taken at zero metres above bed could be real or due to the tip of the sampler dipping down into the bottom sediments while being lowered to the bottom. The samples taken as part of the repeatability test show fairly consistent mass concentration results.

Grain-size analysis was performed on a subset of the suspended sediment samples. The majority of these samples were from the vicinity of Site ES. The sediment collected in the suspended sediment samples closely resembles the bottom sediment samples with a mean ranging from 0.18 - 0.24 mm. The coarser samples were collected closer to the bottom and were less well sorted. The suspended sediment sample from Sites NE and WR, collected with the sampler resting on the bed, have a finer mean grain size than the bottom sediment samples from those sites.

4.2.7   Summary of available measurements

In addition to the USGS field measurement campaign, a wide array of other data are available for calibration and validation of numerical models of Willapa Bay. These include:

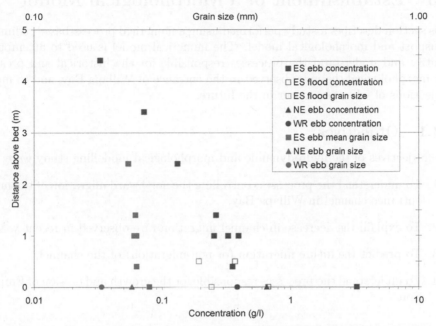

Figure 4.13 – *Results of suspended sediment analysis showing the progressive decrease in concentration and grain size with distance from bottom. Variations in flood and ebb concentrations and grain sizes are also shown. The solid orange squares at 1 meter above the bottom are the samples collected as part of the repeatability test.*

- Long-term measurements of tidal elevations from the NOAA tide gauge located at Toke Point.

- Nearly 10 years of hourly directional wave spectra recorded offshore at the nearby CDIP Grays Harbor wave buoy.

- Process measurements from the earlier 1998 USACE field campaign.

- Process measurements at the mouths of the adjacent estuaries of Grays Harbor and the Columbia River from further USACE field experiments.

- Bathymetric surveys of the bay entrance and main channels have been conducted approximately annually for more than a century and are available from the USACE .

- High-resolution airborne LiDAR surveys of intertidal areas within the bay have recently been conducted by NOAA.

In summary, the USGS field campaign conducted as part of this study added targeted data to an already impressive array of data describing the coastal processes occurring in and around Willapa Bay. The large array of data available at this location makes Willapa Bay an ideal location to develop and test numerical coastal morphological models.

4.3   Establishment of a Morphological Model

This section describes a study performed using a numerical process-based sediment transport and morphological model. The numerical model is used to attempt to identify and understand the processes responsible for the historical and present day morphological changes observed in the entrance of Willapa Bay, and to make projections of what may occur in the future.

4.3.1   Objectives

The objectives of the hydrodynamic and morphological modelling study were:

1. To understand the processes controlling the northward migration of the main entrance channel in Willapa Bay;

2. To explain the decrease in channel migration rate observed in recent years;

3. To predict the future migration (or non-migration) of the channel;

4. To understand the processes responsible for the growth and erosion of Empire Spit.

4.3.2   Modelling approach

Limitations of modelling

Process-based numerical models attempt to simulate the physics of nature by using mathematical formulations describing each of the main physical processes relevant to a particular problem. These mathematical formulations are the result of many years of careful scientific research, however there is much that is still not understood about the physics of complex estuarine and coastal environments. This fundamental shortcoming in the state of scientific knowledge means that no model of sediment transport in a complex estuary such as Willapa Bay can hope to perfectly represent all of the physics involved in morphological change. Even if the physics involved were perfectly understood it would still be beyond the realms of possibility to initialize such a model with the precise conditions occurring at the start of a simulation, or to completely prescribe the fluctuations in forcing conditions that occur during the simulation itself.

The complexity of the real world means that any model is just that, a simplified 'model' of reality. It will not behave exactly as reality does, but hopefully its simplicity assists understanding which processes and process interactions have an important impact on a problem, so that the underlying causes of the problem can be understood. When this process understanding is combined with knowledge of other physical factors that may have a bearing on the problem at hand it can be used to make predictions about possible future scenarios.

Approach

To achieve the modelling objectives outlined above, a range of short- and long-term model simulations were required. In order to validate the sediment transport and morphological model, simulations must be of sufficient duration for measurable changes in channel geometry to occur. For these simulations a target of simulating five years of morphological development was set. Analysing the detail of sediment transport patterns around Empire Spit required higher-resolution simulations averaging the wave conditions occurring over one year.

In order to meet these various objectives, a total of three numerical models were developed. The first model, the regional scale ORegon/WAshington (ORWA2) model, was constructed to bring astronomical ocean tides into the mouth of Willapa Bay. This model was not intended to accurately represent the tidal flows in Willapa Bay, Grays Harbor, or the Columbia River Estuary, but it should accurately capture the propagation of the tide along the continental shelf offshore of Oregon and Washington. Development of this regional model enabled a 'nested' modelling approach. The advantage of this approach is that the smaller, 'nested' models could take advantage of accurate tidal boundary conditions provided by the ORWA2 model. Nested modelling allowed the smaller models to be efficiently designed to focus on Willapa Bay itself. Figure 4.14 shows the ORWA2 model grid and the extent of the nested WBAY and WFINE models.

For modelling hydrodynamic and sediment transport processes in Willapa Bay itself, two local models were nested in the ORWA2 model. The first model, the WBAY model, was designed to model the hydrodynamics, sand transport, and morphological change in the mouth of Willapa Bay over a period of several years. As the objective of the WBAY model was longer-duration morphological simulations, the resolution of the model was kept reasonably coarse. The model resolution is just sufficient to resolve the changing patterns of the main entrance channels and shoals. Unfortunately this means that it has insufficient resolution to resolve detailed sand transport patterns around Empire Spit. The WBAY model was designed to run starting from both 1936 and 1998 bathymetries, in order that historical and present-day sediment transport pathways in the bay entrance could be compared.

The second local Willapa Bay model, the WFINE model, had exactly the same horizontal extent as the WBAY model; however this model had a much higher resolution across the bay entrance and in the vicinity of the main entrance channel and Empire Spit and North Cove. This model ran at approximately one quarter of the speed of the WBAY model and was used for simulating detailed sediment transport patterns in the main channel and around Empire Spit.

4.3.3   Model description

Each of the models was constructed using the Delft3D modelling system including the WAVE and FLOW modules incorporating the 'online' sediment transport and morphological change formulations described in Chapter 2 of this thesis. For the ORWA2 model only the Delft3D-FLOW module was necessary to simulate the propagation of the tide from the deep ocean to the mouth of Willapa Bay. The

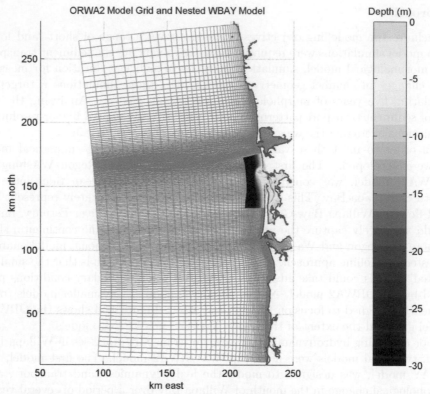

Figure 4.14 *– Regional ORWA2 model grid showing extent of nested WBAY and WFINE models.*

model was applied in depth-integrated (2DH) mode.

The WBAY and WFINE models each make use of both the Delft3D-FLOW and the Delft3D-WAVE modules. The close integration between these modules is essential for modelling the waves, currents, and sediment transport in Willapa Bay because a) the large swell waves breaking on the entrance shoals drive considerable currents in the bay entrance, b) the daily tidal fluctuations in water level have a very significant impact on the penetration of ocean swell waves into Willapa Bay, and c) the tidal currents flowing in and out of the mouth of Willapa Bay may have a significant impact on the waves propagating into the bay. Thus the WAVE and FLOW modules must share a considerable amount of information in order to successfully simulate the interaction of waves and currents in the bay entrance.

Within the Delft3D-WAVE module this study makes use of the third generation SWAN wave model, primarily because the SWAN model is capable of running on the same curvilinear computational grids as Delft3D-FLOW. This considerably simplifies the process of model construction. SWAN is a spectral wave model based on the wave action balance equation. SWAN computes the spatial variation in wave energy that occurs due to generation of wave energy (by wind), propagation of wave energy over water of varying depths, and dissipation of wave energy (primarily by wave breaking). Complete details of the SWAN wave model are presented by Booij et al. (1999) and Ris et al. (1999). In this study, SWAN was run in a 'quasi-stationary' manner in order to simulate temporally changing boundary conditions and water depths. This was achieved by re-running SWAN very frequently (typically after every 10 minutes of hydrodynamic simulation) with updated boundary conditions and water depths, but starting from a 'hotstart' file which captures the wave field at the end of the previous SWAN simulation. The number of iterations SWAN was allowed to perform to converge to a new stationary solution was investigated, but was generally limited to just one iteration, as discussed on Page 91.

The Delft3D-FLOW module with the 'online morphology' option consists of either a 2DH or 3D hydrodynamic model, which solves the continuity and momentum equations for water flow (under the hydrostatic pressure assumption). Integrated within the hydrodynamic model is a sediment transport model that solves the advection-diffusion equation for suspended sediment and a bed-load sediment transport equation. Bed updating due to divergence in sediment transport occurs at every computational time step. The Delft3D-FLOW module is described in more detail in Chapter 2. When the Delft3D -FLOW and -WAVE modules are combined they are capable of simulating most processes relevant to the morphodynamics of coastal systems.

4.3.4   Model setup and forcing processes

ORWA2 model

The ORWA2 model was developed to provide tidal boundary conditions to smaller, more detailed, models of the estuaries in the Pacific Northwest. It achieves this by spanning the region from the deep ocean (150 km offshore) into the shoreline, and approximately 250 km along the coast of Oregon and Washington. The model includes a coarse representation of the estuaries of Grays Harbor, Willapa Bay, and the Mouth of the Columbia River. As this model is only used to simulate the propagation of the large-scale tidal wave along the coast, short waves (ocean swell and wind-generated) are not included in this model.

The ORWA2 model uses the 'curvilinear' grid shown in Figure 4.14. This type of grid requires that grid cells are only approximately rectangular and approximately the same size as neighbouring cells. This flexibility compared to a true rectangular grid allows the construction of flexible and computationally efficient computational grids that can follow the shape of gently curving shorelines. In the case of the ORWA2 model, grid resolution increases from approximately 3-7 km offshore to 500 m or better at the shoreline and inside the estuaries. The grid contains 12,800 active grid cells.

The bathymetry for the ORWA2 model (Figure 4.15) is derived from data presented by Gibbs et al. (2000). The dataset used (ORWA98_750SP) is a digital bathymetric surface grid of the seafloor off the coast of Washington and Oregon. The surface grid was produced by merging NOS Hydrographic Surveys conducted between 1926 and 1974 (primarily 1926 to 1930) with recent surveys conducted by the USACE in 1998 around the entrances to Grays Harbor, Willapa Bay, and the Columbia River. This work represents the best available estimate of the seafloor morphology for the time period circa 1998.

For this study the ORWA2 model is only subjected to tidal forcing. Tidal water level constituents in the deep ocean are obtained from the GOT00.2 global tidal model, which is a refinement of the GOT99.2 model (Ray, 1999). These models, constructed by NASA, combine high-precision satellite altimetry with numerical hydrodynamic modelling to provide the eight main astronomical tidal water-level constituents for the majority of the oceans and seas of the world. Additional important model parameters specified for the ORWA2 model are indicated in Table 4.2. The tidal water level constituents applied to the outer boundary of the ORWA2 model are listed in Table 4.3.

Table 4.2 – ORWA2 model parameters.

Parameter	Unit	Value
Bottom roughness	-	Manning n = 0.025
Horizontal viscosity	m²/s	1
Computational time step	s	60
Latitude	°	46.660
Timezone	-	GMT

Table 4.3 – Tidal constituents applied to the south-west (SW) and north-west (NW) corners of the western boundary of the ORWA2 model.

Tidal Constituent	Amplitude SW (m)	Phase SW (°)	Amplitude NW (m)	Phase NW (°)
M2	0.85	228.4	0.91	234.5
K1	0.42	236.2	0.43	240.0
O1	0.26	220.8	0.26	224.6
S2	0.24	255.9	0.26	263.2
N2	0.18	204.0	0.19	210.2
P1	0.13	233.7	0.13	238.5
K2	0.06	248.8	0.07	256.1
Q1	0.05	226	0.05	229.1
NU2	0.04	203	0.04	209.2
J1	0.03	245	0.03	248
2N2	0.03	195	0.03	201
L2	0.03	237	0.03	243
T2	0.01	244.3	0.02	251.6

WBAY model

The curvilinear grid used for the WBAY model (Figure 4.16) is focused on the entrance of Willapa Bay, however the model includes the entire estuary and the first four kilometres of the Willapa River. The grid extends approximately 15 km offshore to the 50 m depth contour. In the bay entrance the grid has a resolution

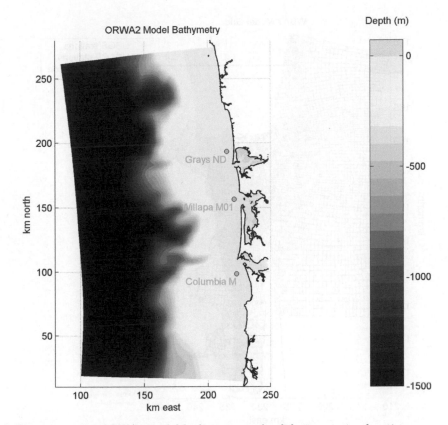

Figure 4.15 – ORWA2 model bathymetry and validation station locations.

of approximately 200 - 300 m, which is sufficient to resolve the main features of
the currents, waves, and sediment transport in the main tidal channels and over
the entrance shoals. Further away from the entrance the grid resolution decreases
to approximately 700 m. The grid resolution has intentionally been kept to the
minimum necessary to resolve the dominant processes in the bay mouth in order to
allow long-duration morphodynamic simulations to be completed at an acceptable
computational cost. The grid contains 9,800 active grid cells.

The WBAY model is run on two different bathymetries. The 1998 bathymetry
(Figure 4.17) is constructed from a number of sources:

- 1998 USACE channel condition bathymetric surveys
- 1998 LiDAR data collected on the entrance shoals
- 1926 NOS hydrographic surveys of the shelf areas off Willapa Bay
- 2002 LiDAR data collected on the inter-tidal areas within the bay

The 1936 bathymetry (Figure 4.18) was obtained by digitizing a historical chart
from the archives of the USACE Seattle district office (Figure 4.34).

The digitized data from the 1936 chart covers most of the entrance to Willapa
Bay and these historical depths were 'pasted' into the surrounding 1998 bathymetry,

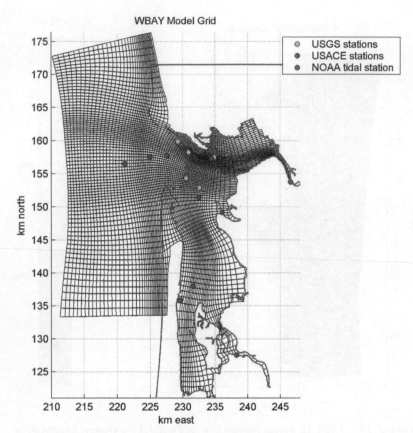

Figure 4.16 – WBAY model grid and station locations.

with transitions between the historical and recent bathymetries estimated where necessary. This makes the assumption that the level and geometry of the mudflats inside Willapa Bay and the deep-water profile of the outer coast have not changed significantly in the last 60 years.

The following processes drive the WBAY model:

Tide   For calibration and validation of the hydrodynamics, tidal forcing is obtained from the regional ORWA2 model by imposing computed water levels on the western boundary of the WBAY model. The northern and southern model boundaries have zero cross-boundary water-level gradient, or 'Neuman' boundary conditions specified.

For longer-duration morphological simulations the tidal forcing is simplified to a mean 'morphological tide'. This simple, repeating, tide has a period of exactly 12 hours 25 minutes and an amplitude of 0.99 m. The morphological tide is chosen to have a morphological impact as close as possible to the average morphological change occurring during an entire neap-spring cycle of real tides. Because of the non-linearity of sediment transport and morphological change, the morphological tide is approximately 8% larger than the 'average' real tide. The selection of the

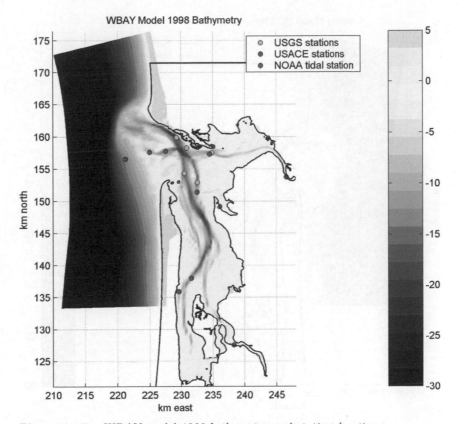

Figure 4.17 – *WBAY model 1998 bathymetry and station locations.*

morphological tide is discussed in more detail on Page 99.

An example time series of actual, astronomical, and equivalent 'morphological' tidal elevations is shown in Figure 4.19. Note that the differences between the actual measured water levels and those predicted using astronomical constituents are due to the 'storm surge' effects caused by the non-tidal processes discussed below.

Waves   Only swell waves arriving from the deep ocean are included in the WBAY model. Swell waves arriving at the outer boundary of the model are obtained from the Coastal Data Information Program (CDIP) Grays Harbor buoy located in 40 m of water approximately 16 km north of the mouth of Willapa Bay. Directional wave data has been recorded every 30 or 60 minutes at this buoy (with a few short gaps) since August 1993. This data provides time series that are applied directly to the model's offshore boundary for simulating the effect of particular storms. The full 10 years of data have also been analysed to provide an average wave climate and to quantify the variations in wave conditions that occur from summer to winter and from one year to another. An example time series of wave conditions recorded at the Grays Harbor buoy is shown in Figure 4.5. The fluctuation of monthly average wave height and direction that occur at the Grays Harbor buoy is shown

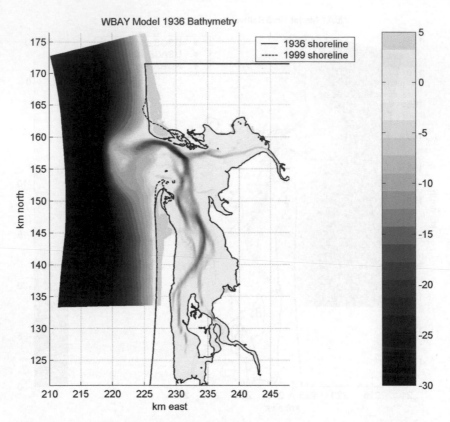

Figure 4.18 – WBAY model 1936 bathymetry.

in Figure 4.20.

Local waves generated by wind are likely to influence sediment transport on shallow areas inside the estuary in areas sheltered from the high-energy swell waves. This effect is neglected in the WBAY morphological model. This approach is adopted as the area of interest is the bay entrance where offshore swells dominate. If the study area was further inside the bay, and more sheltered from offshore swell, then a different method of schematizing the environmental conditions would be required for long-duration morphological simulations.

Other forcing processes   Other forcing processes affecting Willapa Bay include wind, atmospheric pressure fluctuations (and other low-frequency water level fluctuations) and river discharge. These forcing processes were not applied to the long-term morphological simulations performed with the WBAY model as part of this study. The impact of these forcing processes on short term-residual circulation and sediment transport patterns was investigated using the WFINE model described below. The impact of these forcing processes on long-term morphological simulations is discussed in Chapter 5.

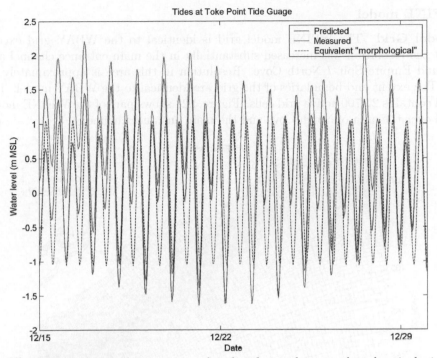

Figure 4.19 – *Example time series of predicted, actual measured, and equivalent morphological tidal water levels.*

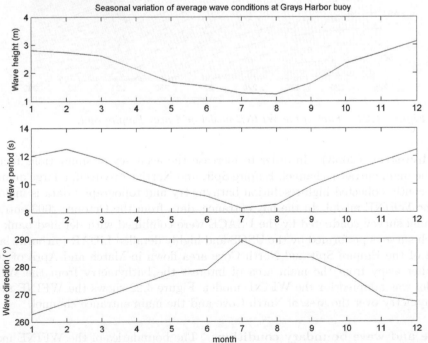

Figure 4.20 – *Monthly average wave height, period, and direction recorded at the Grays Harbor Buoy during 1993-2003.*

WFINE model

Model Grid   The WFINE model grid is identical to the WBAY grid except that the resolution was increased substantially in the main entrance channel and around Empire Spit / North Cove. Resolution in this area is approximately 50 m. The extent and boundaries of the grid are identical to the WBAY model. The grid contains 22,700 active grid cells. Figure 4.21 shows part of the WFINE model grid over the area of North Cove and the main entrance channel.

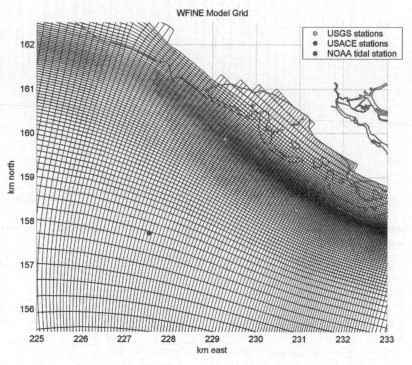

Figure 4.21 – Part of the WFINE model grid near Empire Spit.

Bathymetry (2003)   In order to increase the accuracy of simulations focused on the main entrance channel, Empire Spit, and North Cove itself, a large quantity of recently collected high-resolution bathymetry and topography data is included in the WFINE model. In the bay entrance, data from the October 2003 entrance channel survey conducted by the USACE were combined with detailed bank and beach surveys performed by the USGS and highly detailed LiDAR (airborne laser) data of the Empire Spit and North Cove area flown in March and April of 2002. Further away from the main area of interest the bathymetry from the WBAY model was adopted for the WFINE model. Figure 4.22 shows the WFINE model bathymetry over the area of North Cove and the main entrance channel.

Tide and wave boundary conditions   The boundaries of the WFINE model are forced by the same set of tidal and wave boundary conditions as the WBAY

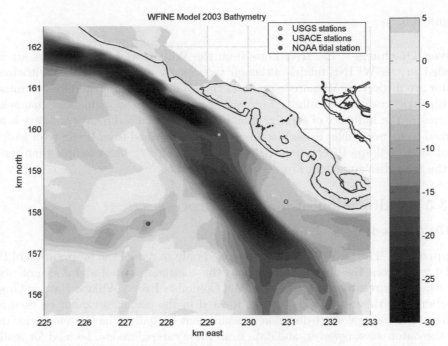

Figure 4.22 – Part of the WFINE 2003 model bathymetry near Empire Spit.

model described above. For the purposes of comparison with measurements, the additional forcing processes of wind, atmospheric pressure fluctuations, and river discharge were also included in the WFINE model.

Wind   Hourly wind data recorded at the Columbia River Bar buoy (NOAA buoy 46029) were used in the WFINE model. In the hydrodynamic model, wind shear stresses generate along-coast flows and across-shelf water-level gradients. Within Willapa Bay, wind stresses cause water-level gradients and circulation currents.

Atmospheric Pressure   Fluctuations in atmospheric pressure cause local elevation and depression of the sea level. As the area affected by one of these pressure systems is much larger than the extent of either the WBAY or ORWA2 models, it is not possible to accurately model the dynamic effects of the moving pressure systems themselves. The time series of atmospheric pressures recorded at the Toke Point tidal station are included however by assuming that this same pressure is acting over the entire WBAY model. This is achieved by including the 'inverse barometer' effect of pressure fluctuations on the model's water-level boundary conditions. This approach neglects the transformation that a large-scale pressure-generated wave will undergo as it approaches the shore; however these second-order effects are minor when compared to the first-order inverse barometer water level effect and the effects of other processes (wind and waves) on the mean water level and currents occurring near the shoreline. An example time series of barometric pressures recorded at the Toke Point tide station are shown in

Figure 4.5.

River Discharge   The river flows from the Willapa and Naselle rivers are included in the WFINE model, although the density effects associated with fresh water are neglected. Measurements show that Willapa Bay is generally well mixed (with little gradient in salinity from surface to bed of the main tidal channels). Therefore the influence of stratification on currents and sediment transport is assumed to be negligible. River discharge data are obtained from USGS river gauges 12013500 (Willapa River) and 12010000 (Naselle River). An example time series of the river discharge data are shown in Figure 4.5.

4.3.5   Calibration of modelled processes

ORWA2 model

Approach   The ORWA2 model is intended only to model the propagation of the tidal wave from the deep ocean up onto the continental shelf and does not need to accurately resolve the estuaries of the Columbia River, Willapa Bay, or Grays Harbor. For this reason tidal gauges located in the estuaries cannot be used for calibration. Several short-duration measurement campaigns have been carried out on the outer coast however, and data from these campaigns can be used for model calibration. For the purposes of calibrating the ORWA2 model the following three campaigns were selected:

1. USACE at Columbia River (April 1998 to March 1999)

2. USACE at Willapa Bay (August 1998 to June 1999)

3. USGS at Grays Harbor (October to December 1999)

All three campaigns used instruments mounted on tripods placed on the seabed to measure waves, currents, and water depth. As currents on the continental shelf are often dominated by wind, and the ORWA2 model is only used to provide water level boundary conditions to nested models the model was only calibrated on the water levels recorded during the field campaigns. One measurement location was selected from each campaign. These locations are shown in Figure 4.15.

Time series of water levels were obtained from the tripod-mounted pressure sensors deployed during the field campaigns listed above. Although an accurate instrument elevation is generally not known, if known atmospheric pressure fluctuations and obvious periods of tripod settling are removed from the record and the mean water level during the campaign is assumed equal to mean sea level, an accurate water level time series can then be extracted from the pressure data.

In order to carry out the calibration the ORWA2 model was run over the duration of each of the field campaigns and a computed time series of water levels was extracted at each of the measurement locations. The computed water levels were then be compared with the measured water levels, and small adjustments to the model boundary conditions were made until satisfactory agreement was achieved.

Table 4.4 – *Calibration adjustments applied to water level boundary conditions in ORWA2 model.*

Tidal Constituent	Amplitude correction factor SW corner (-)	Phase correction SW corner (°)	Amplitude correction factor NW corner (-)	Phase correction NW corner (°)
M2	0.98	+0.0	1.00	+0.0
K1	1.01	+1.0	1.01	+2.0
O1	1.00	+6.0	1.00	−2.5
S2	0.99	+0.5	0.99	+0.5
N2	0.94	−1.5	0.94	−1.5

Table 4.5 – *ORWA2 model, error between measured and computed water levels.*

Campaign / Station	Start date (mm/dd/yyyy)	End date (mm/dd/yyyy)	RMS error (m)
USGS Grays ND	10/5/1999	11/20/1999	0.05
USACE Willapa M01	08/30/1998	10/01/1998	0.10
USACE Columbia M	04/20/1998	06/01/1998	0.14

Calibration adjustments The astronomic water level constituents obtained from the GOT00.2 global tidal model were adjusted slightly to improve the agreement between model and measurements. No other model parameters were changed. The corrections applied to the astronomical water level constituents at the western boundary of the ORWA2 model are shown in Table 4.4.

Calibration results Measured and computed water levels for part of the US-ACE measurement campaign at Willapa Bay are shown in Figure 4.23. Error statistics for the model over each of the campaigns at Grays Harbor, Willapa Bay, and the Columbia River are shown in Table 4.5. A better understanding of the origin of the error in the model results can be obtained from conducting tidal analyses of the modelled and observed water levels at each of the stations. Results for the Willapa M01 station are typical and are presented in Table 4.6.

WBAY and WFINE models

Available data As the WBAY and WFINE models are process-based, measurement data can be used to check the calibration of each of the main processes. This requires observation of tidal hydrodynamics (water levels and currents), waves, and sediment transport. Data collected during two major field campaigns were available to calibrate and validate the WBAY and WFINE numerical models. The first campaign was carried out by the USACE from August 1998 to June

Table 4.6 – *ORWA2 model, error in astronomical tidal constituents at station Willapa M01.*

Constituent	Comp. Amp. (m)	Obs. Amp. (m)	Amp. Ratio (-)	Comp. Phase (deg.)	Obs. Phase (deg.)	Phase Diff. (deg.)
M2	0.942	0.925	1.02	230.6	230.2	+0.4
K1	0.434	0.429	1.01	234.1	236.1	−2.0
S2	0.269	0.264	1.02	257.8	253.8	+4.0
O1	0.253	0.263	0.96	221.4	224.1	−3
N2	0.183	0.188	0.97	194.7	206.1	−11
P1	0.131	0.130	0.99	230.6	232.6	−2.0
K2	0.071	0.069	1.03	250.7	246.7	+4.0
Q1	0.047	0.055	0.85	213.8	219.2	−5

1999. During this campaign several instrumented tripods were deployed to record currents, waves, and water levels, along with a number of land-based tidal water level stations. A meteorological station recording wind speed and barometric pressure was also deployed. Full details of the USACE field campaign are available in Kraus et al. (2000) and data from this campaign were downloaded from http://sandbar.wes.army.mil/.

Although the USACE data provides an excellent basis for calibrating a numerical model of the whole of Willapa Bay, the present study is focused on the mouth of the bay, and the main entrance channel as far as the tip of Toke Point. In order to obtain measurement data more focused on this area, and to collect sediment concentration data which were not collected during the USACE field campaign, the USGS performed a targeted field campaign at the outset of this study, as summarised in Section 4.2.

Approach   The WBAY model was first constructed with a 1998 bathymetry and the tidal hydrodynamics and waves were initially calibrated against water level and current velocity data collected by the USACE in 1998. This model was validated by comparing computed and measured morphological changes. Later, the WFINE model was constructed with a 2003 bathymetry and identical boundary conditions to the WBAY model. The WFINE model relied on the calibration of the WBAY model, and was validated against the process data collected by the USGS in 2002-3.

Due to the high wave climate it was only possible to collect hydrodynamic and sediment transport data in the main estuary channels, and not on the broad entrance shoals. The model shows, however, that waves breaking on the entrance shoals play an important role in determining the residual sediment transport and currents in the mouth of the bay. It would therefore be highly desirable to gather process data on the entrance shoals in order to check the calibration of these processes in an environment very different to that found in the deep channels. This, unfortunately, remains one of the greatest weaknesses of this modelling study and could only be remedied by a future field measurement campaign carefully targeted at capturing the hydrodynamic and sediment transport processes occurring on the entrance shoals.

The model validation consisted of two 50 day simulations covering the duration of the second USGS field deployment (7 December 2002 to 19 January 2003). The first simulation was performed with only tidal forcing applied to the boundaries of the WFINE model in order to check the propagation of the tide from deep water in to the Toke Point tide station. The second simulation included all forcing processes recorded during the field deployment (atmospheric pressure, wind offshore waves, and river discharge). This allowed validation of the non-tidal components of the model.

Tidal water levels   Figure 4.24 shows measured and computed tidal water levels at the Toke Point tidal station. In this plot the measured water levels have been filtered to remove the effects of non-tidal forcing as the model is driven by tidal forcing only. The modelled water levels achieve a correlation R=0.99 and an RMS error of 0.087 m.

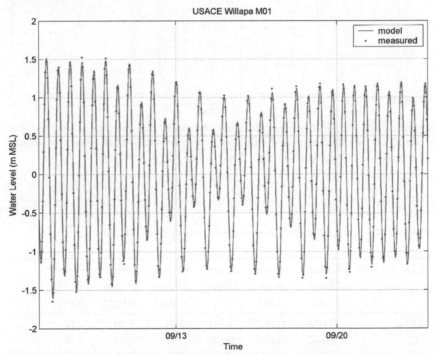

Figure 4.23 – *ORWA2 model validation. Measured (dots) and computed (solid line) water levels (m MSL) at USACE Station Willapa M01. Time is mm/dd in 1998*

The goodness of fit of the computed water levels at Toke Point were further analysed by performing a tidal analysis on both the computed and observed water levels over the simulation duration. Table 4.7 compares computed and observed tidal constituents.

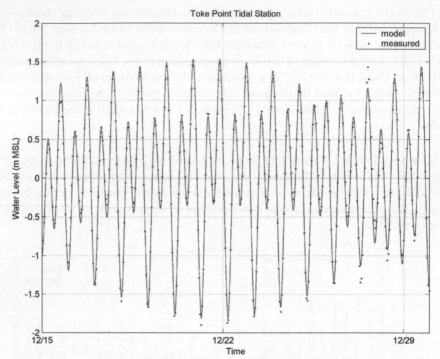

Figure 4.24 *– WFINE model validation. Measured and computed water levels at Toke Point tide station. Non-tidal frequencies were removed from measurements. The model is driven by tide only. Time is mm/dd in 2002*

Table 4.7 *– Computed and observed tidal constituents at Toke Point.*

Constituent	Comp. Amp. (m)	Obs. Amp. (m)	Amp. Ratio (-)	Comp. Phase (deg.)	Obs. Phase (deg.)	Phase Diff. (deg.)
M2	0.958	0.957	1.00	256	256	0
K1	0.565	0.586	0.96	259	259	0
O1	0.284	0.287	0.99	233	225	+8
S2	0.201	0.196	1.02	286	282	+4
N2	0.152	0.164	0.93	223	225	−2
Q1	0.052	0.037	1.41	235	204	+31
J1	0.042	0.043	0.98	264	249	+15

In general, the measured and computed water level components compare very well, with amplitudes within 5% and phases within 10 degrees. The small components Q1 and J1 do not compare particularly well, however there is a large degree of uncertainty in the tidal analyses of these components due to their small amplitude and the relatively short analysis period.

Total water level   When non-tidal forcing is included, the temporal variation of water levels at Toke Point is more complex than the tide-only situation considered above. Figure 4.25 shows measured and computed water levels at Toke Point over a 2-week interval during the second USGS field deployment. In this plot, the model includes all forcing processes and the measurements are unfiltered. The deep water model boundary was raised by 0.20 m to take into account regional water-level effects that are not included in the WFINE and ORWA2 models, thereby ensuring that the mean water level computed over the total duration of the second USGS deployment matched the mean water level measured at Toke Point.

Careful inspection of Figure 4.25 shows that the amount of storm surge caused by non-tidal processes fluctuated from more than 1 m on December 16th down to as little as 0.15 m on December 23. The model captures this complex temporal variation in non-tidal storm surge well. The correlation between computed and measured total (unfiltered) water levels is again extremely high with R=0.99. RMS error increases to 0.15 m however this is good for unfiltered water levels over such a stormy period.

Waves   Wave heights were measured during the second USGS deployment at Station ES on the north-eastern flank of the main channel adjacent to Empire Spit. The wave heights at this location are small relative to the height of the waves on the outer coast, however they provide a useful check of the reduction in wave height occurring over the entrance shoals.

During the second USGS deployment, offshore wave heights fluctuated from approximately 1 m to 6.5 m, however wave heights at Station ES fluctuated from approximately 0.2 m to around 1 m. Figure 4.26 shows wave heights offshore and at Station ES for the 2-week period from 15th to 30th December. The tidal modulation of the wave height (high waves at high tide, low waves at low tide) at Station ES is very clear in both the data and the model results, however the model does appear to consistently over-estimate the wave height at the station, especially during times of high offshore waves. This could be due to a number of factors, including the quality of the available bathymetric data on the entrance shoals themselves. In absolute terms, the RMS error between the modelled and measured wave heights at Station ES is only 0.24 m. This will likely result in a very slight (insignificant) overestimation of computed sediment concentrations and longshore sand transport in the shallow water adjacent to Empire Spit.

When used in conjunction with the "online morphology" version of Delft3D-FLOW, SWAN is run in a "quasi-nonstationary" manner, this involves running stationary SWAN simulations frequently (every 10 to 15 minutes of hydrodynamic time) with updated boundary conditions but starting from the final state of the previous SWAN simulation. SWAN is generally not allowed to fully converge to

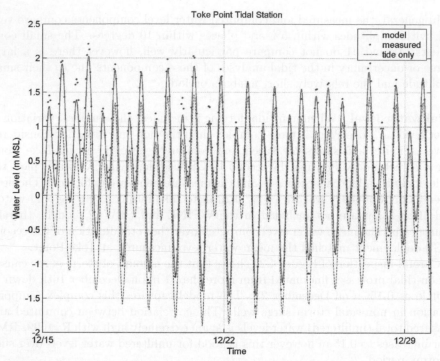

Figure 4.25 *– Measured and computed water levels at Toke Point tide station.*
Measurements unfiltered, model includes all forcing processes. Time is mm/dd
in 2002.

the new solution, but is terminated after just a few iterations. In this manner
the SWAN solution is always approaching, but never reaching, a new stationary
solution. This is not a recommended method of running SWAN and a better
solution would be to more closely link the two models so that SWAN could run
fully non-stationary simulations and dynamically exchange updated wave, current,
water and bed elevation data with Delft3D-FLOW.

In order to verify the performance of the SWAN model in use at Willapa
Bay a number of tests were performed where the SWAN model was allowed a
varying number of iterations to reach convergence at each call. The tests were
performed over the USACE and USGS instrument deployments in 1998 and 2002
respectively and included simulations which both included and excluded the effects
of currents. The RMS error and correlation coefficient for the SWAN model for
various numbers of iterations are shown in Figure 4.27. These results confirm that
the RMS error in the SWAN model is lowest when current is included and just one
iteration is allowed. This result is likely to be the consequence of other calibration
settings that were determined using simulations with just one iteration in order to
save computational effort. At station ES the correlation coefficient is slightly lower
when 3 or 5 iterations are allowed. At USACE#2 the pattern is opposite, with 3 or
5 iterations yielding a slightly higher correlation coefficient. On the basis of these
results the SWAN model is run with just one iteration. Using the standard SWAN
convergence criteria this results in convergence being reached in approximately

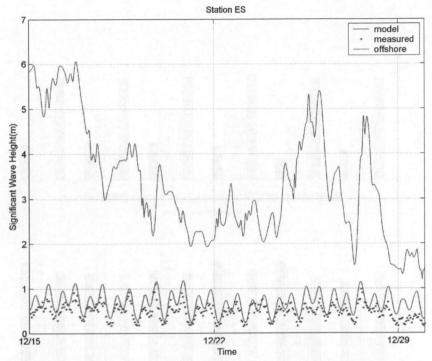

Figure 4.26 *– Measured and computed wave heights at Station ES over a stormy 2-week period. Time is mm/dd in 2002.*

90% of grid cells, although this fluctuates between approximately 55% and 95% depending on stage of the tide and changes in offshore wave boundary condition.

Currents   At the five USGS stations (refer to Figure 4.4) instruments measured profiles of current velocity during each deployment. These data were subsequently depth integrated by fitting a logarithmic velocity profile based on an assumed bed roughness from the bed to the bottom bin of the profiler, and assuming a constant velocity from the top bin of the profile to the water surface. A principal component analysis was then performed on each time series of depth-integrated velocities in order to determine along- and across-channel directions at each site. Table 4.8 shows the measured and computed principal flood directions at each station.

As the flows in the channels are largely parallel to the channel axes, the along-channel current velocity components (positive in the flood direction) of the measurements can be compared to the model. Figure 4.28 shows measured and modelled time series of along-channel velocities at stations ES and MC for the 5-day period from 15th to the 20th December 2002. Correlations for the entire second USGS deployment are shown in Figure 4.29. Results at other stations are not shown here, but are similar. A summary of the main characteristics of the velocity correlations at all stations is presented in Table 4.9.

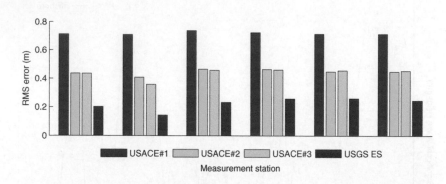

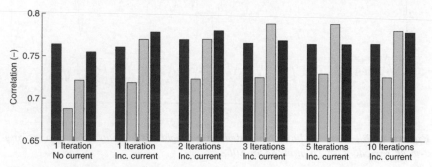

Figure 4.27 – *Error and correlation between SWAN wave model and measurements for varying SWAN iteration settings.*

Table 4.8 – *Measured and computed principal current directions.*

Station	Observed principal flood direction (deg.)	Computed principal flood direction (deg.)	Directional error (deg.)
ES	129	132	+3
WR	86	89	+3
NW	174	173	−1
NE	186	183	−3
MC	128	136	+8

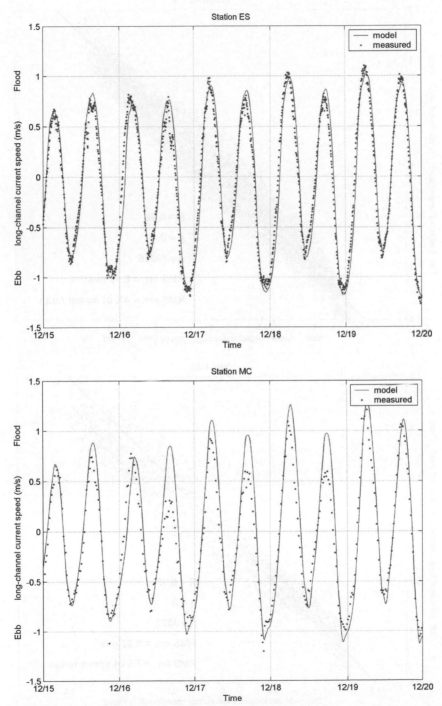

Figure 4.28 *– Measured and computed unfiltered along-channel current velocities at Station ES (upper panel) and Station MC (lower panel).*

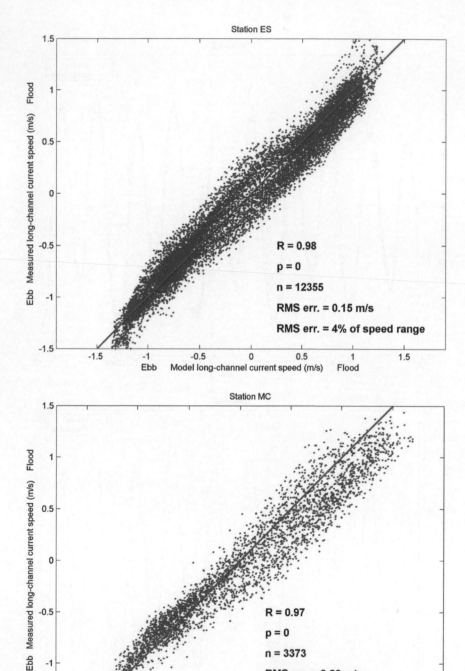

Figure 4.29 *– Correlation between measured and computed unfiltered long-channel current velocities at Station ES (upper panel) and Station MC (lower panel).*

Table 4.9 – *Comparison of measured and modelled current velocities.*

Station	RMS error (m/s)	Percent error	Qualitative observations
ES	0.15	4	Systematic under-prediction of both peak ebb and flood velocities. Slight phase error visible.
WR	0.15	5	Significant under-prediction of peak ebb velocities.
NW	0.14	5	Excellent agreement. Ebb-directional tidal asymmetry accurately captured.
NE	0.14	5	Excellent agreement. Flood-directional tidal asymmetry accurately captured.
MC	0.22	7	Better agreement on ebb than flood. Tends to over-predict flood velocities.

Table 4.10 – *Measured and computed residual current velocities.*

Station	Measured residual speed (direction) (m/s)	Computed residual speed (direction) (m/s)
ES	0.05 (ebb)	0.05 (ebb)
WR	0.04 (flood)	0.00
NW	0.06 (ebb)	0.07 (ebb)
NE	0.14 (flood)	0.12 (flood)
MC	0.05 (ebb)	0.03 (flood)

Residual currents   Averaging the measured and modelled along-channel current velocities over time results in small, but important, residual along-channel velocities at each station. These residual velocities are important as they also tend to indicate the direction of residual sand transport at each station. The measured and computed long-channel residual velocities over the duration of the second USGS deployment are presented in Table 4.10. The results are generally very good. Stations ES, NW, and NE show excellent agreement between computed and measured residual currents. The results at WR and MC are not so good, with the model predicting a small flood-directed residual velocity at Station MC when a small ebb-directed residual was measured. This discrepancy may be due to local bathymetric effects, or changes in bathymetry between survey date and instrument deployment. It may also indicate that the model generally under estimates the ebb dominance of velocities in the main channel, possibly because of underestimating the wave-driven circulation of water in the estuary entrance.

Suspended sediments   It is difficult to obtain accurate suspended sediment concentration data, especially in deep channels with high current velocities. Figure 4.30 shows a collection of sediment concentration measurements obtained from optical backscatter (OBS) and direct sampling techniques at Station ES. As waves are relatively unimportant at Station ES, the measured concentrations are plotted against depth-integrated current speed. Measurement data are presented for two heights above bed (0.46 m and 0.74 m) for a range of current speeds. Superimposed on the OBS data are a small number of manual suspended sediment samples that were taken at the same station, as well as the suspended sediment concentrations predicted by the model.

Although a lot of scatter is visible in the OBS measurement data, the model results do seem to form a reasonable lower bound to the OBS measurements, and fit the manual suspended sediment samples rather well. Considering the uncertainties present in the measurements, and the sediment pick-up formulations used in the model, this is a remarkable level of agreement. It is clear that modelled and measured suspended sediment concentrations show the same trends and are in the same order of magnitude.

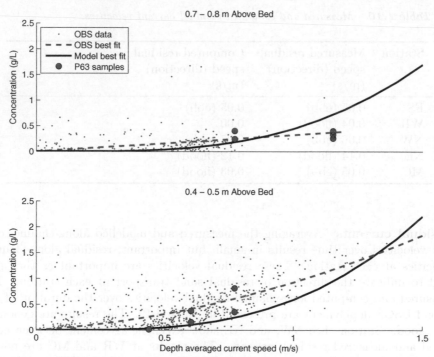

Figure 4.30 *– Measured and computed suspended sediment concentrations at Station ES.*

4.3.6   Morphological model setup

Although the morphological model is process based, ultimately one of the main modelling objectives is to reproduce and predict the morphological changes that are observed in nature. In general these morphological changes are determined not by any one process but rather by a complex balance between several processes. To this end, comparison of computed and measured morphological changes helps to understand to what extent the model accurately captures this interaction between multiple processes, and to what extent its morphological predictions can be relied upon.

Approach

Two distinct recent periods of morphological change have been recorded in Willapa Bay. For much of the last century the main entrance channel was migrating rapidly to the north. However, in recent years the position of the channel appears to have stabilized. The morphological model was tested by running medium-duration (3-5 year) morphological simulations during each of these phases to see if the model could reproduce the observed morphological responses. The two periods chosen were 1936 to 1939 when the channel was migrating rapidly, and 1998 to 2003 when the channel position was more stable.

In order to simulate morphological change over a period of 3-5 years a number of acceleration techniques must be used in order for the simulations to be completed within a realistic time frame. Without use of these acceleration techniques even the efficiently designed WBAY morphological model, running on the fastest personal computers (PCs) available today, only runs at approximately 60 times faster than real time, and thus a 5-year simulation would take around 30 days to compute. By utilizing the input reduction and acceleration techniques described below the speed of the WBAY morphological model is increased to almost 3000 times real time, and a 5-year morphological simulation can be completed in 15 hours. A detailed investigation and refinement of the following morphological acceleration techniques is presented in Chapter 5.

Morphological tide selection

A single 'morphological tide' was selected and applied as the deep-water boundary condition. This single tide is a simple sinusoid with a period of 12 hours 25 minutes (approximately the same period as the main M_2 tidal constituent) and amplitude of 0.99 m (8% larger than the actual M_2 constituent). In reality the M_2 tidal constituent represents the average tidal range and $M_2 + S_2$ (amplitude $0.88 + 0.24 = 1.12$ m) produces an average spring tide while $M_2 - S_2$ (amplitude $0.88 - 0.24 = 0.64$ m) produces an average neap tide. In the Pacific Northwest the diurnal constituents K_1 (0.42 m) and O_1 (0.26 m) are also rather large, resulting in a very 'mixed' tide and tidal ranges are frequently significantly larger than the average figures given above. Theoretical work by van de Kreeke and Robaczewska (1993) has shown that for the purposes of medium term morphological simulations these 'odd' constituents can be neglected because the manner in which they interact with the main semi-diurnal constituents results in zero net flux of sediment. This finding is investigated in more detail in Chapter 5.

Previous studies at Delft Hydraulics (unpublished) using the method described by Latteux (1995) to establish a morphological tide for studies in several regions of the world have confirmed Latteux's finding that the selected morphological tide is generally 7-20% larger than the M_2 constituent alone. This finding was locally reconfirmed by Gelfenbaum et al. (2003) in their study of the adjacent Grays

Harbor estuary mouth when they used Delft Hydraulics' "Latteux" procedure to select a morphological tide of 1.084 times the main M_2 constituent.

This study simply followed Gelfenbaum et al. and adopted a morphological tide of 1.08 times the offshore M_2 component. Sensitivity tests were conducted to check the sensitivity of the morphological simulations to the choice of the morphological tide. Qualitative comparisons of the results of these tests reveal that the computed morphological response of the estuary entrance is insensitive to the precise size of the chosen morphological tide, compare Figure 4.31 ($1.2 \times M_2$), Figure 4.32 ($1.08 \times M_2$), and Figure 4.33 ($1.0 \times M_2$). A more detailed investigation of the morphological tide for Willapa Bay is presented in Chapter 5.

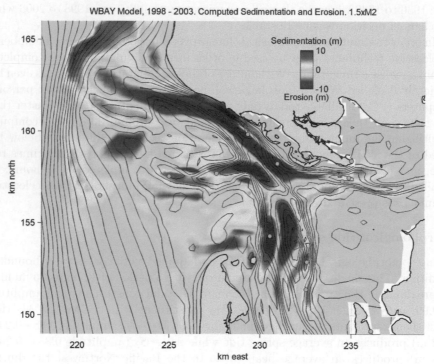

Figure 4.31 – Computed morphological change 1998–2003 using a morphological tide of $1.2 \times M_2$.

Wave climate schematization

The wave climate in the north-eastern Pacific Ocean is characterized by summer conditions with smaller waves approaching the coast from the north-west and winter conditions with larger, longer period, waves approaching the coast from the south-west.

For the 5-year morphological simulations covering the period from 1998 to 2003 the actual chronology of wave events is known and this is schematized as 10 successive "seasons". Each season is six months long, beginning with "winter 1998/99" which covers the period 1 November 1998 to 30 April 1999. The waves

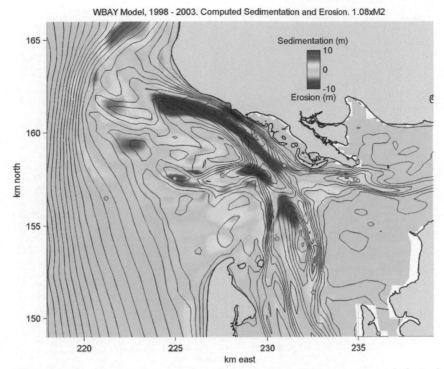

Figure 4.32 – *Computed morphological change 1998–2003 using a morphological tide of* $1.08 \times M_2$.

occurring each winter season are divided into 12 discrete classes of wave height and direction with associated probabilities of occurrence. Due to the smaller range in wave conditions and lower overall morphological impact, summer seasons are divided into 7 classes, resulting in a total of 19 wave classes representing all the wave conditions occurring in a year. During the classification process within each height and directional bin each wave condition is assigned a weight proportional to H_s raised to the power of 2.5 in order to give an estimate of the morphological significance of the wave condition. The resulting schematised wave climates used for each of the 10 seasons are located in Appendix A. Note that the winter of 1998/99 was part of an El Niño cycle and the higher wave energy occurring during this winter required a different arrangement of the wave class boundaries than in subsequent winters. The tables in Appendix A also show a representative wind for each wave class. These were not used in the simulations described in this chapter, but were used in the simulations described in Chapter 5.

For the 3-year morphological simulation from 1936 to 1939 and for future projection (2003 to 2008) simulations the actual chronology of wave events or real inter-annual wave climate fluctuations is unknown. For these simulations it is assumed that the typical fluctuation between summer and winter conditions is still valid, however all years are assumed to have the same, average, distribution of wave energy. In these cases a repeating two-season wave climate computed from the full 10-year 1993 to 2003 wave record (Table 4.11) is used. Within each season

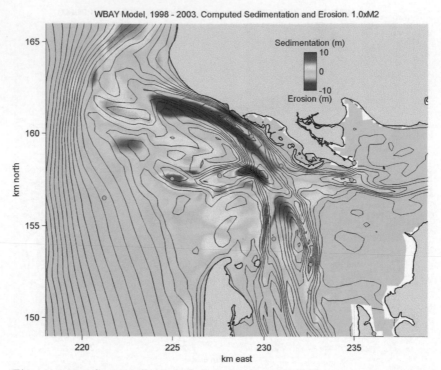

Figure 4.33 – Computed morphological change 1998–2003 using a morphological tide of $1.0 \times M_2$.

the order of the separate wave classes is randomly assigned.

Morpholgical acceleration factor

The morphological acceleration factor (morfac) is a technique used to bridge the gap between hydrodynamic and morphological timescales. At each hydrodynamic time step both the fluxes of suspended sediment to and from the bed and the bed-load sediment transport components are multiplied by morfac. If morfac is greater than 1.0 then the morphological time step is effectively made longer than the hydrodynamic time step.

During a morphological simulation, each of the selected wave conditions is simulated for the duration of one morphological tide (12 hours 25 minutes = 745 mins) in order to account for the random phasing between waves and tides that occurs in nature. Morfac is then used to increase the morphological changes occurring during this one tide to the changes that would occur during the entire duration of that wave condition in one year. By using this approach, 19 morphological tides (one for each wave condition) are required in order to simulate one morphological year. This implies that an average morphological acceleration factor of $365 \times 24 \times 60/(19 \times 745) = 37.13$ is applied to the simulation overall. For each wave condition the morphological acceleration factor applied will depend on the percentage occurrence of that particular wave condition. For example, for a

Table 4.11 – *Average two season wave climate schematisation.*

(a) Winter (duration 181 days).

Wave bin (°)	Representative direction (°)	Rep. H_s (m)	Rep. Period (s)	Dur. (%)	Dur. (days)	Morfac (-)
$H_s < 1.2m$						
180 - 360	279	1.00	10.2	10	18	35.34
$1.2m \leq H_s < 3.0m$						
180 - 230	216	2.28	8.6	4	7	13.05
230 - 250	241	2.29	10.0	4	7	14.37
250 - 270	262	2.16	11.4	9	17	31.92
270 - 285	278	2.16	13.0	20	36	69.96
285 - 360	294	2.06	11.4	21	38	72.75
$3.0m \leq H_s < 5.0m$						
180 - 240	224	3.88	10.1	6	11	21.18
240 - 270	257	3.87	12.2	7	13	25.93
270 - 285	278	3.79	13.9	9	17	32.48
285 - 360	291	3.71	13.4	5	10	18.77
$5.0m \leq H_s < 9.0m$						
180 - 270	241	5.76	12.9	2	4	8.17
270 - 360	281	5.90	15.1	2	3	5.78

(b) Summer (duration 184 days).

Wave bin (°)	Representative direction (°)	Rep. H_s (m)	Rep. Period (s)	Dur. (%)	Dur. (days)	Morfac (-)
$H_s < 0.8m$						
180 - 360	271	0.70	9.8	6	11	21.32
$0.8m \leq H_s < 2.5m$						
180 - 240	221	1.67	8.4	5	10	18.85
240 - 270	259	1.58	9.9	12	22	42.03
270 - 290	281	1.64	10.4	28	51	98.97
290 - 360	300	1.53	8.6	33	60	116.79
$2.5m \leq H_s < 9.0m$						
180 - 270	241	3.50	10.7	7	12	23.16
270 - 360	284	3.38	13.2	10	18	34.61

common wave condition that occurs for 20 days per year the morphological acceleration factor used would be $20 \times 24 \times 60/745 = 38.66$. For a large (rare) wave condition that only occurs for 12 hours in a year the morphological acceleration factor used would be $12 \times 60/745 = 0.966$.

This approach has the desirable effect that higher acceleration factors are applied to the more common, and generally smaller, wave conditions during which the morphology is less active, and smaller acceleration factors are applied to the larger (and less common) wave conditions (when the morphology is more active and large acceleration factors might cause a problem). The morphological acceleration factors applied to each wave condition of each season are indicated in Table 4.11 and Tables A.1 to A.10.

4.3.7   Morphological model sensitivity and calibration

Approach

Bathymetries were digitised for the years 1936, 1939, 1998, and 2003, with the early bathymetries obtained by digitizing USACE nautical charts (Figures 4.34 and 4.35). The 1998 and 2003 bathymetries were constructed from modern bathymetric and LiDAR survey data. By differencing bed elevations, patterns of sedimentation and erosion can be identified. These patterns are used for comparing computed and measured morphological trends.

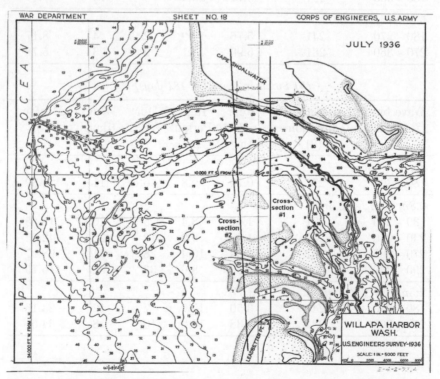

Figure 4.34 – 1936 bathymetry map. Depths in feet.

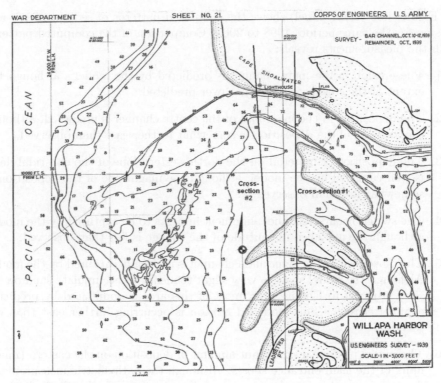

Figure 4.35 – *1939 bathymetry map. Depths in feet.*

Morphological changes 1998 - 2003

Figure 4.36 (upper panel) shows the patterns of erosion (blue) and accretion (red) that were measured during the period 1998-2003. White areas did not have repeated survey data available. Several trends are clear in this plot:

1. The outer half of the entrance channel became significantly deeper.

2. A large amount of sediment was deposited in the inner end of the centre channel.

3. A large quantity of material was deposited on the north-west lobe of the ebb-tidal delta.

4. The outer (western) end of the main channel (west of approximately 225 km east) appears to have switched from an alignment that runs almost due west, to a more northerly alignment. Sedimentation is visible in the old outer channel, and significant erosion is occurring to the north.

5. The inner main channel (east of the SR105 groin) appears to be moving away from the northern edge of the channel. Deposition is apparent on the outer edge of the shelf immediately adjacent to Empire Spit, and significant erosion was observed on the southern side of the channel.

Figure 4.36 (lower panel) shows the computed patterns of sedimentation and erosion for the same period, 1998 to 2003. Comparison of the computed patterns with the measurements reveals:

1. The erosion in the main channel is predicted by the model - although the extent of the erosion is somewhat over predicted.

2. The deposition in the inner end of the centre channel is predicted, although the location of the deposition is somewhat further east than observed.

3. The deposition of material on the north-western lobe of the ebb-tidal delta is well captured by the model, although the magnitude of the deposition is somewhat less than observed.

4. The switch in outer channel alignment appears to be captured by the model, although the modelled magnitude of the change is less than observed.

5. The movement of the inner main channel is not really captured by the model. Although the model shows a very slight tendency to accumulate sand on the shelf adjacent to Empire Spit, most of this area is dominated by predicted channel erosion. This modelled erosion is occurring further east than observed.

6. The model predicts a significant amount of deposition on the eastern (inner) flank of the main entrance shoals. Although the bathymetric measurements do not extend this far, the amount of deposition seems unrealistically high. In longer (10-year) simulations this deposition can completely block the Nahcotta West channel, and this does not occur in nature.

Morphological changes 1936 - 1939

Figure 4.37 (upper panel) shows the observed patterns erosion and accretion during the period 1936 - 1939. The following patterns are apparent:

1. The northward migration of the main channel is clear. Erosion occurs along the north edge of the channel between about 225 and 230 km east. Sedimentation occurs adjacent to this on the southern side of the channel.

2. Deposition occurs along the north edge of the outer main channel (from approximately 221 to 225 km east, 157 km north) and in an area of old channel at 222 km east, 155 km north.

3. Infilling and northward migration of a small secondary channel crossing the entrance shoals at approximately 153 km north.

4. Deposition on the eastern side of the main entrance shoals.

The computed patterns of sedimentation and erosion for this period are shown in Figure 4.37 (lower panel). Comparing the computed patterns with the measured patterns reveals:

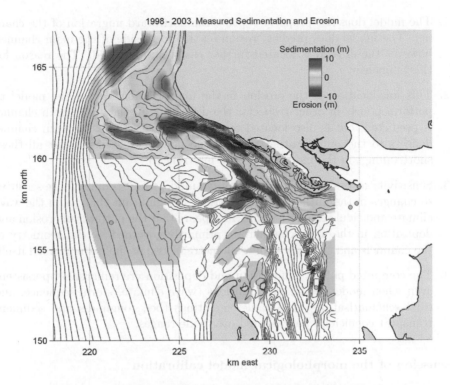

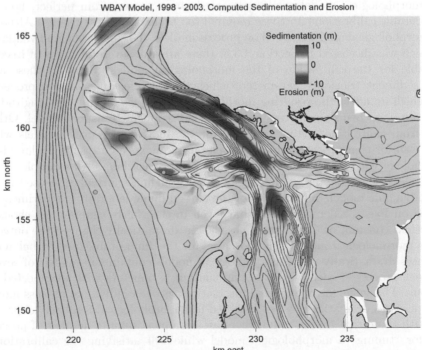

Figure 4.36 – *Measured (upper panel) and computed (lower panel) sedimentation and erosion 1998 to 2003 (shown on 1998 bathymetric contours).*

1. The model does not properly capture the northward migration of the channel. The model does predict erosion on the north side of the main channel, however the erosion is located further east than observed. The reason for this is unclear.

2. This mis-location of the erosion in the main channel causes the model to perform poorly in other respects: the deposition in the outer main channel is predicted, but is also located too far to the east, and the main channel deflects to the south, rather than to the north. It is likely that all these shortcomings in the model result are interrelated.

3. Sensitivity tests show that the computed model result is much more sensitive to changes in the initial bathymetry than to significant changes in the wave climate and tidal schematization. It appears that the patterns of erosion and deposition in the main channel are primarily governed by the geometry of the channels and shoals within a kilometre or two of the main channel itself.

4. The computed patterns of erosion and deposition are remarkably persistent even when model boundary conditions (wave climate, storm sequence, and tidal schematization) or parameter settings (bed roughness and sediment transport coefficients) are altered over wide ranges.

Discussion of the morphological model calibration

The morphological model results presented above are far from perfect, however considerable calibration effort was required to obtain even these results. Although the morphological model is based on process models, the calibration and validation of which was discussed in Section 4.3.5, there are still several degrees of freedom available for the calibration of the morphological model. Some of these arise from sediment transport parameters which do not affect measurable processes, but which do affect the resulting morphology. Examples include the longitudinal and transverse bed slope factors in the bedload sediment transport model. Others arise from calibration settings selected for the individual process models which work well for that process model, but may have unexpected results on long-term morphology when applied in a morphological feedback loop. An example of this latter class of parameter is the choice of bed roughness formulation used in the hydrodynamic model. Simulations of the curved flume laboratory experiment reported on Page 38 demonstrated that, even in that fairly simple case, choice of equivalent constant or Chezy roughness height dramatically changed the outcome of the morphological model. The potential for this kind of unexpected behaviour increases dramatically when morphological models contain sub-models of several different processes, each of which may contain parameters with unexpected behaviour. The scope for unexpected interaction between individual process models also increases dramatically when included in a morphological feedback loop.

With those thoughts in mind, there are an almost infinite number of possibilities for "tuning" a morphological model while still satisfying the calibration of each of the underlying process models. A huge amount of work can also be incurred. For example, changing the formulation used for calculating the enhanced

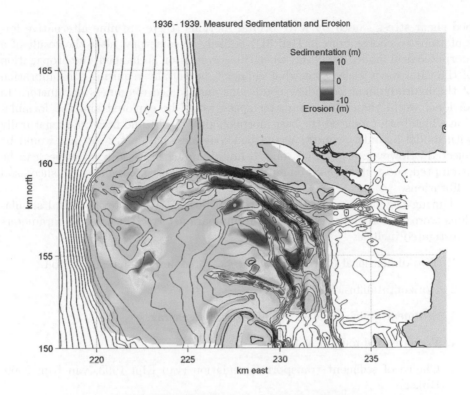

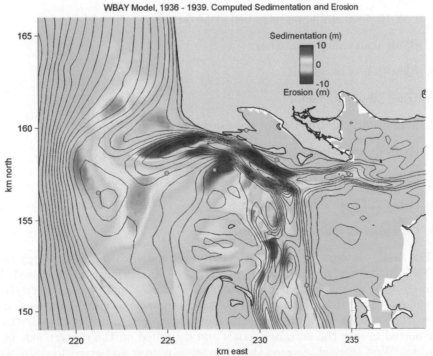

Figure 4.37 – *Measured (upper panel) and computed (lower panel) sedimentation and erosion 1936 to 1939 (shown on 1936 bathymetric contours).*

bed shear stress caused by wave orbital motions (there are nine alternative formulations to choose from in Delft3D) is likely to somewhat alter the result of a
morphological model. However, when this formulation is changed, the propagation
of the tidal wave is also somewhat changed, potentially requiring a recalibration
of the hydrodynamic boundary conditions and/or bed roughness parameter. In
an ideal world the entire parameter space would be explored and the formulation which both produced the best morphological results and fitted most naturally
with model boundary condition and bed roughness parameter settings would be
selected. However, in practice, only a finite number of these possibilities can be
attempted, most likely based on gut instinct or previous experience with successful
calibrations.

During the study reported here, several dozen 5-year morphological simulations were performed investigating various parameter settings. The parameters
investigated include:

1. Two-dimensional (depth averaged) vs. three-dimensional simulation

2. Horizontal diffusion coefficient

3. Sediment grain size available at the bed in channels

4. Wave-related roughness height

5. Choice of sediment transport formulation (van Rijn 1993, van Rijn 2000,
   Bijker)

6. Longitudinal and transverse bed slope factors

7. Bank avalanching algorithm

8. Elevation of the shoals

9. Wave streaming factor

10. Magnitude and direction of the wave climate

11. Hydrodynamic bed roughness formulation (Chezy vs. constant K_s)

12. Tidal phase difference applied to the offshore boundary

13. Details of wave schematisation

14. Choice of morphological tide

The main problems being investigated were related to the morphology of the
entrance channels, the very feature on which the simulations were focussed. The
channels tend to suffer from two problems a) with default parameter settings
the channels tend to develop to become too deep and too narrow relative to the
observed channels and b) the model predicts that large quantities of sediment are
transported in over the entrance shoals and deposited on the western side of the
Nahcotta West channel, causing this channel to narrow and even close altogether
in longer-duration simulations.

The first problem, of the channels becoming too narrow and deep, is somewhat affected by the choice of sediment transport formulation, but is most likely caused by a combination of two factors. First, a model with relatively coarse spatial resolution, such as the WBAY model, is not able to accurately represent the steep gradients in bathymetry occurring at channel edges and, second, it is possible that an additional down-slope sediment transport process, such as some kind of avalanching mechanism, plays an important role on morphological time scales and is entirely missing from the present model. Based on these hypotheses the problem was alleviated by dramatically increasing the long- and cross-flow downslope bedload transport factors. The values used in the final morphological model ($f_{ALPHABS} = 70$ and $f_{ALPHABN} = 100$) are unrealistically high at approximately 70 times larger than the default values. These were the only model parameters which were found to be at all effective in alleviating the problem and it is therefore likely that (ab)use of this model process is being used to cover for another process missing from the model formulations. It is likely that in doing so errors are introduced into other areas of the model.

The second problem, of a lot of sand accumulating on the west side of the Nahcotta West channel, is possibly caused by the down-slope problem described above. However another factor, the rate of sand import across the entrance shoals, could also play a role. Unfortunately, the problem of sand import over the entrance shoals is rather complex and affected by many factors. First, the true elevation of the entrance shoals is poorly known - especially for the historical simulations. Second, breaking waves drive an strong residual current in over the entrance shoals. The magnitude of this current is critical in determining the rate of residual sediment transport and is affected by many parameter settings, from wave breaking parameters to hydrodynamic bed roughness, wave enhancement of bed roughness, etc. As waves tend to break strongly on the entrance shoals it is also an area of intense wave stirring of sediments and possibly transport of sediment by wave asymmetry. It is also possible that wave groups play a role, and these are completely neglected from the present model. In short, considerable uncertainty exists regarding model settings affecting inward sand transport over the entrance shoals and, as no process data exists on the shoals, the only method of calibrating these processes is by observation of the resulting morphological change compared to observed change.

Summary

The following conclusions are drawn from the process of calibrating the morphological model:

1. Qualitatively, the morphological model does a reasonable job of capturing the main morphological changes observed during the 1998 - 2003 period.

2. The 1936 - 1939 simulation displays many of the features observed during this period, but ultimately fails to reproduce the rapid northward channel migration.

3. The cause of the erroneous 1936 model result is unclear. Extensive sensitivity

tests were performed on the model and these revealed that, at a gross scale, the computed patterns are insensitive to moderate changes in model forcing processes and calibration settings, except for the initial bathymetry.

4. As the model performs well when the initial bathymetry and boundary conditions are well known (1998 - 2003) and significantly less well in 1936 - 1939 when the initial bathymetry was less well known, it seems likely that errors in the 1936 bathymetry used as input to the model simulation are the cause of the discrepancies observed in this simulation.

5. It is highly likely that the model is missing a process (such as the slumping of steep channel slopes) that may have a very local, but important, effect on the lateral migration of the main channel.

6. The above observations lead to the conclusion that the sediment transport patterns computed in the 1998 model are reasonably reliable. In the 1936 model the general patterns of sand transport are considered reliable, but the details of the transport computed at the rapidly eroding tip of Cape Shoalwater clearly are not. It appears that the model can only explain causes of channel migration in terms of general patterns of sediment transport.

4.4   Morphological Model Analysis

4.4.1   Analysis of WBAY morphological simulations

Approach

Although the morphological model results discussed in Section 4.3.7 failed to reproduce all the features of the observed morphological changes, careful analysis of the broader patterns of sediment transport predicted by the model may give insight into the reasons for the change in observed behaviour between 1936 and 1998. Comparison with future prediction simulations may also help understand likely future developments. As the absolute model results contain systematic errors, an approach of interpreting the differences between pairs of model simulations was employed. This makes the assumption that differencing model results will tend to cancel systematic errors, leaving a clearer picture of the real differences in behaviour between the two modelled situations. The analysis was approached as follows:

1. Model simulations are performed with selected physical processes disabled in order to identify which processes dominate the transport of sand in the entrance to Willapa Bay.

2. Identical simulations are performed using the 1936 and 1998 bathymetries (as reported in Section 4.3.7). The computed sediment transport and morphological change patterns were contrasted in order to identify important changes between the two periods.

3. Predictive simulations are performed starting from the 2003 bathymetry in order to compare predicted near-future transport and morphological change patterns with the two historical cases.

4. The understanding gained by performing the above simulations was used in conjunction with knowledge of the underlying geology, to make predictions for the future migration of the main channel.

Typical ebb, flood and tidal residual currents

Figure 4.38 shows typical computed flood current patterns for Willapa Bay in 1998 and 1936. Figure 4.39 shows typical ebb current patterns. The simulations have been arranged so that the plots correspond to identical points in an identical tidal cycle. The colour in the plots indicates the speed of the current in metres per second. Depth contours are shown in white. It is clear that the model predicts that both flood and ebb current velocities in the main channel were higher in 1936 than in 1998. The marked curvature of the flow in the main channel in 1936 is also clear. Current velocities are much higher in the main channel than over the shoals. The 1998 simulation shows a significant flow through the secondary (centre) channel. This feature did not exist in 1936, although flow does appear to cross the shoals in a number of more minor secondary channels. Flows inside the estuary appear similar in both simulations.

Plotting the patterns of computed residual tidal currents in Willapa Bay gives an indication of the patterns of net water (and perhaps sediment) circulation that occur in the bay due to tidal currents alone. Figure 4.40 shows the computed residual tidal current patterns for 1998 and 1936.

In 1998, the strongest residual tidal currents are directed out of the estuary along the eastern flank of the entrance shoals and in the centre channel and the outer portion of the main channel. A net inward flow of water occurs over the entrance shoals. This general pattern of water tending to flood in over shoals and ebb through channels is expected for an estuary with large intertidal areas.

In 1936 the residual tidal currents were stronger and tended to converge from the south to a 'pinch point' in the main access channel immediately south of Cape Shoalwater. From this point a strong residual current flows offshore in the main entrance channel. This is an interesting pattern as it is likely that such clear residual tidal velocities would cause similar patterns of net sand transport.

Annual total residual currents and sediment transport patterns

The previous results showed residual currents due only to tidal forcing. When other forcing processes are included (wind, waves, river discharge, and atmospheric pressure fluctuations) additional residual currents occur. Figure 4.41 shows computed annual residual current patterns when additional forcing processes are included in the simulations. In the bay entrance the only other significant contributor to depth averaged residual currents is the effect of waves breaking on the entrance shoals.

When comparing the annual residual current plots with the residuals due to tide alone it is clear that many of the main residual current patterns are unchanged,

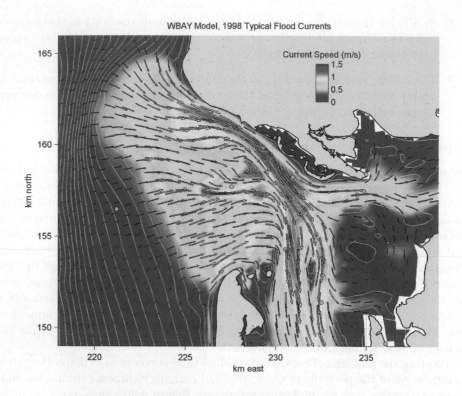

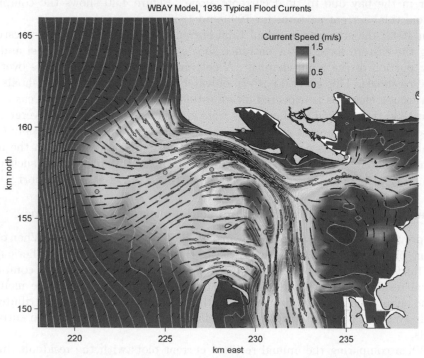

Figure 4.38 – Typical flood currents, 1998 (upper panel) and 1936 (lower panel).

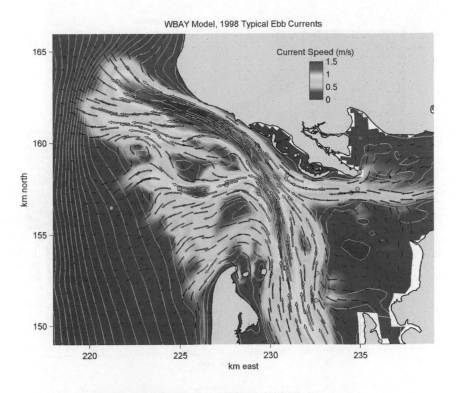

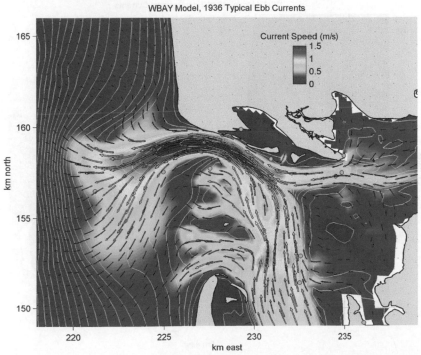

Figure 4.39 – *Typical ebb currents, 1998 (upper panel) and 1936 (lower panel).*

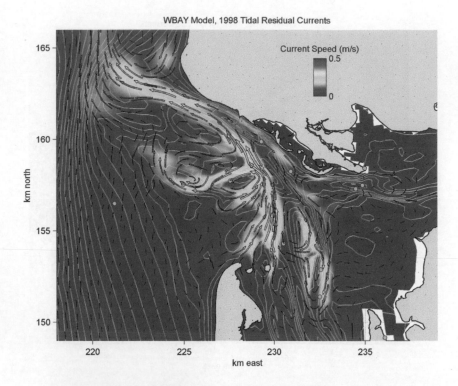

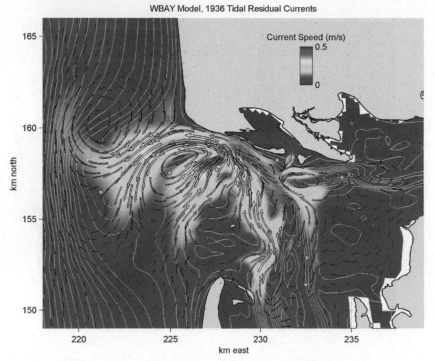

Figure 4.40 – *Residual currents due to tide only, 1998 (upper panel) and 1936 (lower panel).*

however breaking waves tend to drive additional flow in over the entrance shoals and enhance the outward-directed residual flow in the main channel. In addition, breaking waves drive significant residual currents from the outer (western) tip of the ebb-tidal delta in towards the coastline. They also cause localized circulation patterns where water is driven inwards over shallower areas and then returns out through the deeper channels. These wave-induced changes appear to be consistent in both the 1998 and 1936 models.

Figure 4.42 shows the computed annual residual sand transport patterns around the mouth of Willapa Bay in 1998 and 1936. These plots are the result of simulations using the schematized two-season wave climate described earlier (Table 4.11). The simulations were each 19 tidal cycles in duration with a different wave condition occurring for each tidal cycle. Together the 19 tidal cycles make up a schematized morphological year. Figures 4.43 and 4.44 show tidally-averaged sand transport patterns for typical 'winter' and 'summer' wave conditions separately, in order to illustrate the areas of similarity and difference in sand transport that occur under the different wave conditions that combine to produce the annual averages.

Several important features are visible in the residual sand transport plots:

1. All residual sand transport plots, for both the 1998 and 1936 models, show general sand transport patterns over the entrance shoals and in the main channel that are remarkably similar to the annual residual current patterns discussed above.

2. Although the different wave conditions clearly have a significant impact on the magnitude and direction of sand transport on the outer ebb-tidal delta, the general pattern of transport of sand inwards over the shoals and outward through the outer main channel is consistent in all simulations.

3. Residual sand transport rates in and around the main channel were higher in 1936 than in 1998. This is consistent with the higher ebb, flood, and residual currents identified earlier, and the higher rates of morphological change observed historically.

4. In 1936 a strong divergence of sand transport is apparent in the main channel. This will cause strong erosion in the main channel adjacent to Cape Shoalwater. The divergence of sand transport in the main channel in the 1998 model is much less pronounced.

5. In 1936 there was clearly a strong supply of sand to the south side of the 'pinch point' in the main channel, immediately south of Cape Shoalwater. The supply of sand to the south side of the main channel appears to be largely a recirculation of sand carried out of the estuary in the main channel and is remarkably consistent in all wave conditions in 1936. In 1998 there is no obvious supply of sand to feed the shoal on the south side of the main channel.

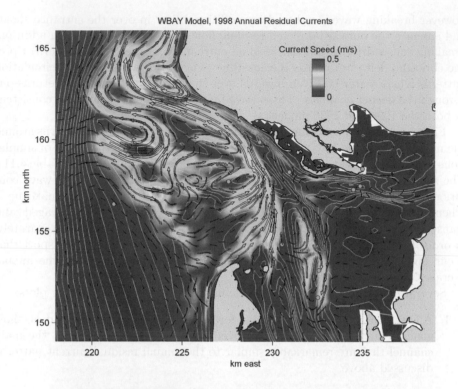

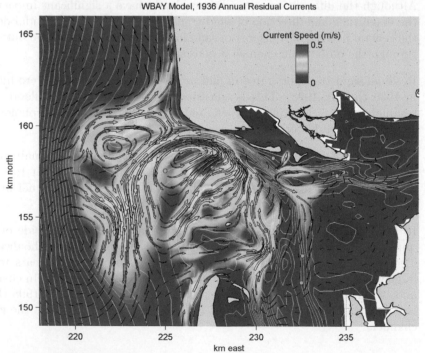

Figure 4.41 – *Residual currents due to tide, wind, and waves, 1998 (upper panel) and 1936 (lower panel).*

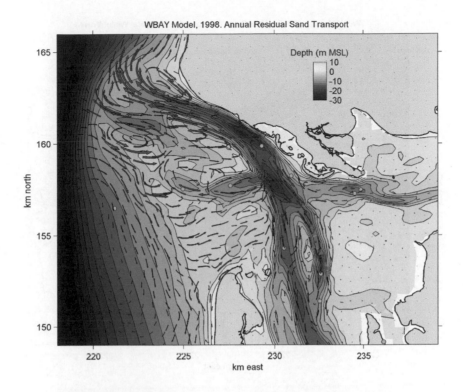

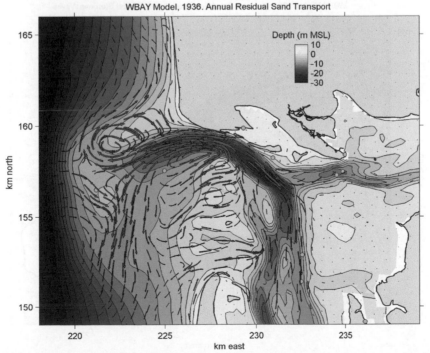

Figure 4.42 – *Annual residual sand transport, 1998 (upper panel) and 1936 (lower panel).*

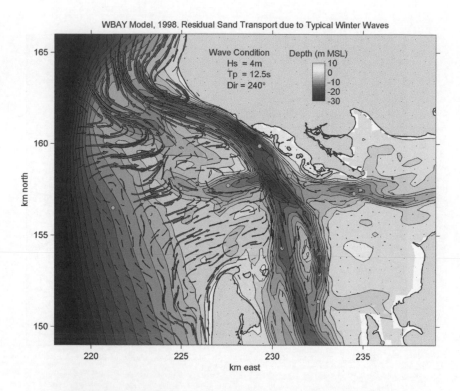

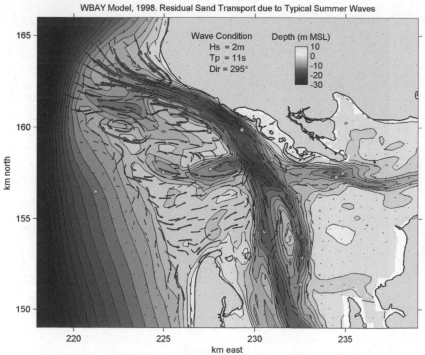

Figure 4.43 – *Tide-averaged sand transport, Typical winter (upper panel) and summer (lower panel) conditions, 1998.*

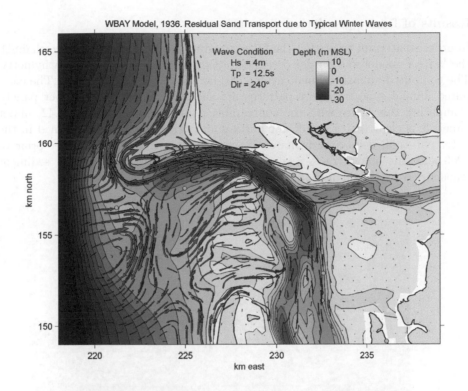

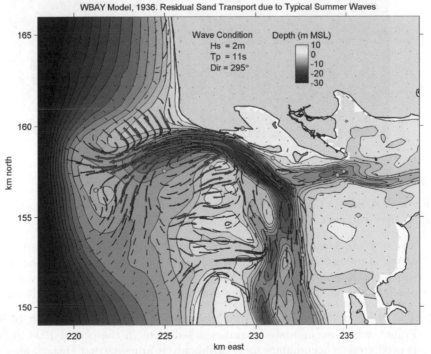

Figure 4.44 – *Tide-averaged sand transport, Typical winter (upper panel) and summer (lower panel) conditions, 1936.*

Results of future prediction simulations

In order to attempt to predict the future migration of the main entrance channel, the WBAY model was run for 5 years, starting from the latest 2003 bathymetry. The computed annual residual current pattern is shown in Figure 4.45. The computed annual sand transport patterns are shown in Figure 4.46 (upper panel). Comparing this figure with the comparable figure for 1998 (Figure 4.42, upper panel) it is clear that a significant north–south orientated channel formed in the outer ebb-tidal delta approximately 2 km offshore of Cape Shoalwater prior to 2003 and this now attracts a considerable proportion of the flow and sediment transport from the main channel.

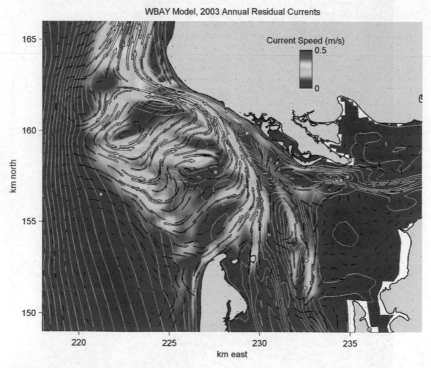

Figure 4.45 – 2003, annual residual currents (all forcing processes).

Figure 4.47 shows the patterns of sedimentation and erosion that are predicted to occur over the next 5 years. Notable features of this result are:

1. The new northward extension to the main channel offshore from Cape Shoalwater is predicted to accumulate sediment and erosion is predicted further south on the ebb-tidal delta. This pattern appears to be caused by a switch of the outer main channel from a northerly direction to a southerly one.

2. The old centre channel (cutting through the entrance shoal west of Toke Point), which accumulated significant sediment in the period 1998 to 2003, is predicted to accumulate more sediment. It appears that this channel may completely fill up and become extinct in coming years.

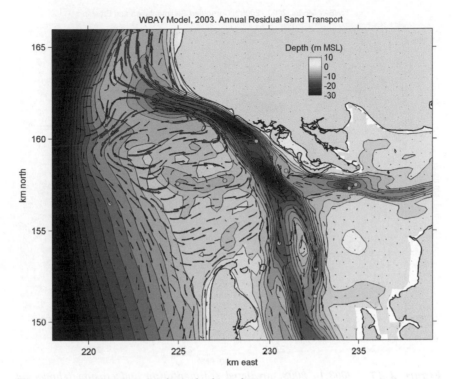

Figure 4.46 – 2003, annual residual sand transport.

3. Significant accumulation of sand is predicted in deep water at approximately the location of USACE station #1 (Figure 4.5). This is associated with the predicted switch of the main channel to a more southerly orientation. If this switch occurs then it appears that deposition to the southern lobe of the ebb-tidal delta is likely to recommence.

4. The massive accumulation of sand predicted on the eastern side of the entrance shoals is a common feature of all simulations. It appears that the model over predicts sand accumulations in this area. At this point in time the reason for this inaccuracy in the model is not known (refer to the discussion on Page 111).

4.4.2   Analysis of WFINE annual sediment transport simulations

Observation of the morphology and historical shape of Empire Spit suggests that this feature has formed as a result of littoral (wave-driven) transport along the shallow shelf that exists on the north side of the main channel. As such, it is interesting to use detailed modelling to compute the magnitude and direction of this littoral transport and the interactions it may have with the sand transport that is occurring in the main channel. As with the WBAY model analysis, absolute results from the morphological model were not used directly, rather differences between

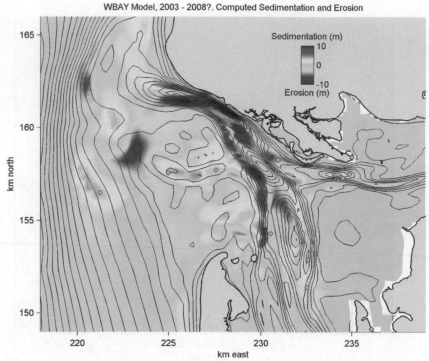

Figure 4.47 – *2003 to 2008, predicted sedimentation and erosion (shown on 2003 bathymetric contours).*

similar model results and trends visible within model results were interpreted to help explain the observed morphological change patterns.

Annual residual sediment transport patterns

Residual sand transport patterns in the main channel and around Empire Spit (Figure 4.48) were obtained by averaging over a year of tide and wave conditions. Analysis of these residual sand transport patterns provides insight into the processes driving morphological change in the main channel and around Empire Spit. Analysis starts by calculating the annual quantity of sand transported through a number of cross sections arranged across the main channel and up on the shelf on the north side of the main channel. The following findings can be drawn from the data presented in Figure 4.48.

Littoral (wave-driven) sand transport on the north side of the main channel   Littoral transport on the shelf on the north edge of the main channel was investigated by performing a 1-year simulation with the WFINE model in which an 'obstacle' was inserted into the wave model along the centreline of the main channel. This obstacle absorbed all wave energy reaching the centreline of the channel, thereby preventing wave-driven transport along the shelf on the north edge of the channel. When the results of this simulation were subtracted from the standard WFINE model simulation the effects of waves could then be separated from other transport mechanisms. Wave-driven littoral transport along the north edge of the main channel is directed into Willapa Bay and steadily decreases from 300,000 m^3/yr near the mouth of the channel to about 8,000 m^3/yr at the south-eastern tip of Empire Spit. This decreasing transport is mainly due to the decreasing wave energy further into the bay.

At three locations along the northern edge of the channel, asymmetries in the tidal currents appear to dominate the effect of waves on transports along the shelf.

1. Immediately east of the SR105 groin, asymmetry in tidal currents (due to acceleration and deflection of the ebb current by the groin) cause a reversal of the direction of the along-shelf transport. This effectively blocks all littoral transport past the SR105 groin.

2. At the two gaps between the islands that make up Empire Spit the shelf tends to widen slightly. This widening deflects the flood tide out into the channel slightly, thereby causing it to accelerate and locally increase sediment transport. It appears that these local areas of tide-dominated transport cause local increases in the littoral transport and play an important role in the ongoing breaching and reshaping of the islands that make up Empire Spit.

It is clear that exchange of sediment between the shelf and main channel must occur in order that continuity of sand mass is maintained. The following exchanges are clear:

1. Between the outer coast and the SR105 groin there must be a progressive transport of sand from the shelf down into the main channel, otherwise massive accretion of sediment would occur west of the SR105 groin.

2. Transport of sediment from the main channel up onto the shelf must occur for 1 - 2 km east of the SR105 groin in order to supply the continuation of the eastward-directed littoral drift east of the groin.

Sand transport in the main channel   Extending the analysis from the shallow shelf on the northern edge of the main channel to include the entire main channel provides the following insights:

1. The littoral transports along the shelf on the north edge of the channel are 2 orders of magnitude (i.e., 100 times) smaller than the residual sand transport occurring in the main channel.

2. Gross ebb sand transport in the main channel is approximately two-times greater than the gross flood transport. Adjacent to the base of Empire Spit (section Main4) average sand transport during ebb is estimated to be 12,000 m^3/tide, flood transport is estimated to be 6,000 m^3/tide. This results in an estimated average outward-directed residual transport of 6,000 m^3/tide or 2.8 million m^3 of sand per year passing out through the main channel at this point.

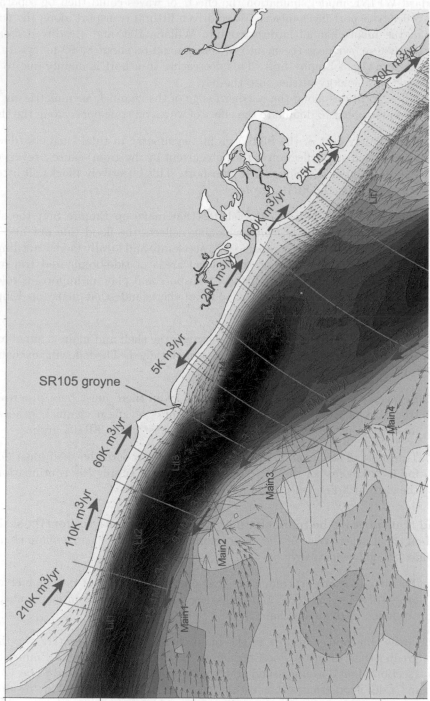

Figure 4.48 *– Detailed annual residual sand transport patterns along the northern coastline and within the main channel of Willapa Bay.*

3. When considering the length of the main channel, from the mouth of the bay to Toke Point and Bay Center, a large gradient in residual sand transport in the main channel exists. Near the mouth the residual export of sand is very strong (estimated 16 million m^3/yr) however the computed residual transport of sand past Toke Point and Bay Center are rather small (approximately 100,000 m^3 import per year). Either sand transported into the bay over the entrance shoals and/or deepening and widening of the main channel must supply this gradient in the main channel. The spatial imbalances between supply and demand for sand will dictate the morphological evolution of the main channel.

4. It is also interesting to examine patterns of sand transport within the main channel. Sand tends to move from the entrance toward Toke Point along the northern side of the channel. This transport pathway carries sand from the deep scour hole in the vicinity of the SR105 groin and deposits it on the broad flat shelf on which USGS Station ES was located. At this point the inward-directed residual transports on the north side of the main channel are opposed by the outward-directed residual transports on the north side of the Willapa River channel (see Figure 4.48). This convergence of residual sand transport is likely to be responsible for the observed stability of the location of the confluence of these channels over the last several decades. The southern side of the main channel shows a strong export-dominated residual transport.

The pronounced asymmetry in sand transport in the main channel is caused primarily by a pronounced asymmetry in current velocities in the main channel. This asymmetry in current velocities can be understood by analyzing the hydrodynamics of the bay entrance. The tidal prism of Willapa Bay (for a 'morphologically average' tide) is approximately 800 million m^3 and approximately 70% of this water (550 million m^3) is exchanged through the main channel. Significantly more water flows outward (ebbs) through the main channel than flows in (floods). This is caused by three processes:

1. A natural tendency for tides to flood over shoals and ebb through channels. Modelling indicates that this effect alone accounts for a residual outward-directed flow of approximately 40 million m^3 per tide in the main channel (7% of the flow in the main channel).

2. Waves breaking on the entrance shoals push water in to the estuary over the shoals. This water flows back out to sea via the deeper main channel where the forces due to breaking waves and resistance to flow are lower. Although the amount of water pushed over the shoals depends on the size of the offshore waves and the tidally-varying depth of the shoals, modelling indicates that on average this effect produces a residual outward-directed flow of approximately 100 million m^3 per tide in the main channel (18% of the flow in the main channel).

3. The Willapa and Naselle Rivers discharge fresh water into the bay and this water exits through the bay mouth to the Pacific Ocean. Peak storm fresh-

water discharge into Willapa Bay can reach 1000 m^3/s, or 45 million m^3 per tide, however these peak river discharges are of short duration (a few days at most). When longer-term averages are considered (which are more important for long-duration morphological development) freshwater input into Willapa Bay is estimated to be 90 m^3/s, or 4 million m^3 per tide (0.7% of the flow in the main channel). This is a relatively insignificant contribution to residual flow in the main channel when compared to the tidal residual and wave-driven flows described above.

As the main channel has a cross-sectional area of approximately 22,000 m^2, the average residual current caused by tidal asymmetry and wave-driven currents amounts to an average outward flow of approximately 10 cm/s. This flow reduces the amount of sediment carried in by the flooding tide, and increases the sediment carried out by the ebbing tide.

4.5   Morphological Modelling Conclusions

4.5.1   Historical trends

Entrance channel migration

Numerical modelling indicates that in historical times, when the entrance channel was migrating rapidly to the north, tidal current velocities and sediment transport in the main channel were higher than they are today. The stronger tidal currents did not in themselves cause the channel to migrate, however they were a symptom of the 'pinching' of the entrance channel adjacent to Cape Shoalwater. Model simulations indicate that the force causing this pinching action was the historical shape of the outer ebb-tidal delta, which tended to transport large quantities of sand to the entrance shoal immediately south of Cape Shoalwater.

Perhaps a useful metaphor for this is to think of the tidal flow in the main Willapa Bay entrance channel as a sharp metal file which is pulled back and forth with each tide, the stronger the currents, the sharper or coarser the file. However, even a sharp file will not cut well unless it is pressed against the object upon which it is placed (in this case Cape Shoalwater). In the past, pressure was applied by the sand deposited on the entrance shoal immediately south of Cape Shoalwater. Because of this applied pressure, both Cape Shoalwater and the entrance shoal to the south would have been progressively 'filed' away. However, the entrance shoal to the south had a steady supply of sand and this kept up the pressure on Cape Shoalwater. In time, Cape Shoalwater simply wore away.

Growth and erosion of Empire Spit

From the numerical modelling it is clear that wave-driven currents on the narrow shelf on the north side of the entrance channel do provide a small but continuous transport of sand in to the estuary along this side of the channel. Because of the decreasing wave energy further into the bay the amount of this wave-driven sand transport steadily decreases along the length Empire Spit. This gradient in sand transport provides a small ongoing source of sand to Empire Spit. However, in its current configuration, this wave-driven supply of sand is estimated to be rather small (less than 50,000 m^3 of sand spread along the length of Empire Spit each year).

When seeking to understand the historical rapid development of Empire Spit it is useful to return to the 'metal file' analogy for the erosion of Cape Shoalwater discussed above. In 1936, when the pressure on the southern side of the 'file' was high, a large amount of material (sand) was filed (eroded) from Cape Shoalwater and the opposing entrance shoal. This eroded material was carried away from the pinch point by the tidal currents, and then deposited. A large amount of sand was carried outwards by ebb tides and deposited on the large submerged spit clearly visible in the 1936 bathymetry (Figure 4.18). Similarly, a significant proportion of the eroded material was carried into the estuary on flood tides. Some of this material was deposited on the edge of the main channel (see Figure 4.42, lower panel) where it was then re-worked by wave action to contribute to the development of Empire Spit. In 1936, this supply of sand to Empire Spit would have been far more significant than the limited wave-driven sand transport along the north side of the main channel.

4.5.2   Present condition

Entrance channel migration

In recent years the shape of the outer ebb-tidal delta has changed, and this has removed much of the sand supply to the shoal on the south side of the main channel. Two main changes have occurred to bring about this change:

1. The entire ebb delta to the south of the main channel has prograded sea-wards. This has resulted in the MSL -6m depth contour (a depth at which wave breaking is very intense and a large amount of sand is transported) having a distinct NW - SE orientation. This geometry prevents much of the sand that is transported northwards along Long Beach Peninsula during winter storms from reaching the south side of the main channel.

2. The outermost reach of the main channel has been in a westerly or northerly alignment since at least 1998. This has meant that the large quantities of sand exported through the main channel have been deposited to the west, north-west, and north of the mouth of the main channel, rather than to the south-west where it is pushed back in onto the entrance shoal immediately south of the main channel.

The end result of these two changes is that in recent years pressure has been taken off the 'file' in the main channel and the northward migration of the entrance channel and consequent erosion of Cape Shoalwater has slowed. In fact, because of the reduced supply of sand into the south side of the main channel, the main channel has 'relaxed' back to the south and increased in width and depth somewhat in recent years.

Growth and erosion of Empire Spit

Because of the reduction in erosion of material from Cape Shoalwater, and the present strong outward asymmetry in sand transport in the main channel, there is little sand being supplied from the main channel to the shallow littoral zone adjacent to Empire Spit. This means that the only remaining supply of sand to maintain Empire Spit is the small wave-driven transport along the north side of the main channel. Furthermore, it is possible that the recent construction of the SR105 groin may have further reduced this small supply of sand from the outer coast. This means that in the present situation any loss of sand from the littoral transport along Empire Spit can only be replaced by erosion of the spit itself. This has manifested itself as thinning, lowering, and eventual overwashing of Empire Spit near the point at which it attaches to the mainland. The sand transported by littoral transport is carried to the south-east along the spit. Some is carried into North Cove where it deposits and remains; some is deposited near the southern end of Empire Spit where it is responsible for the gradual elongation of the spit, and the remainder is directed out into the main channel where it is lost in deep water. Cross-shore processes (overwash and undertow) may also contribute to loss of material from the outer face of Empire Spit.

4.5.3   Why the change?

Entrance channel migration

Following the analysis of historical and present-day sand transport patterns, the migration of the main entrance channel depends critically on the supply of sand to the entrance shoal on the southern side of the main channel. As described above, the reduction in 'pressure' on the southern side of the main channel in recent years has come about because of two distinct changes in the outer ebb-tidal delta. The more frequent of these changes, the 'switching' of the outer reach of the main channel from a northerly to southerly alignment, as described by Hands and Shepsis (1999), is known to occur in a cyclic manner and the next switch to a southerly alignment can be expected to occur at any time. Indeed, future prediction simulations performed as part of this study predict that the channel will undergo a swing to the south within the next 5 years. Although this result is expected, the exact timing of the next switch to a southerly alignment cannot be predicted as it may depend to a large extent on the precise sequence of storm events to act on the outer ebb-tidal delta. In any case, when this southerly switch occurs, a large proportion of the sand carried out through the main channel will then be deposited offshore of the shoal to the south of the entrance channel and, after some time, it is likely that this material will begin to increase the sand supply to the shoal, and thereby begin to apply increased pressure to the south side of the entrance channel.

    The second change in the ebb-tidal delta, the progradation and overall change in alignment of the MSL -6 m depth contour, has taken place over several decades and appears to be a consequence of the northerly rotation of the entrance to the main channel, possibly in combination with the presence of a secondary central channel in recent years. Volume change analysis shows that the total volume of sand contained in the bay entrance (including Cape Shoalwater) has not significantly changed from 1936 to 2003 (Figure 4.49), therefore the change in ebb-tidal delta geometry has occurred primarily by redistribution of sand already contained in the entrance, rather than by addition of large quantities of sand from an external source.

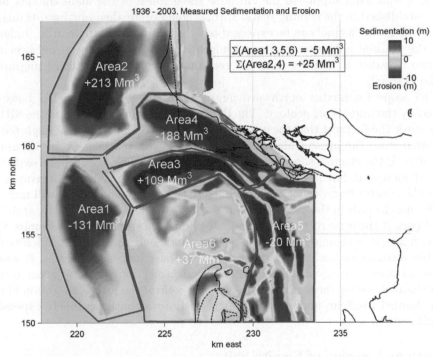

Figure 4.49 *– Measured sedimentation and erosion, 1936 – 2003.*

Growth and erosion of Empire Spit

In the present situation, the supply of sand to Empire Spit is limited to sand transported by tidal currents and deposited on the margins of the main channel, as the construction of the SR105 groin has blocked all littoral sand transport to the spit. The spits do however slowly lose sand both into North Cove and out into the main channel, especially during storm events. The end result of this is that Empire Spit is becoming progressively lower and is slowly 'rolling' backwards over the marshland behind it. These processes can be expected to continue, perhaps at an accelerating rate as the lower spit is more frequently overtopped by storm waves.

4.5.4   Future predictions

Entrance channel migration

Considering the historical dynamics of the ebb-tidal delta, and the dominant processes acting on it today, there is no reason why the outer ebb delta should remain in its existing configuration in the coming decades. There is every likelihood therefore that sand supply to the entrance shoal south of the main channel may be re-established in the coming years. If this should occur then northwards migration of the main channel can be expected to accelerate shortly thereafter. Judging from the gradual reduction in channel migration rate that has occurred over the last two decades, it is likely that any recommencement of northward migration will be gradual.

The scope for further northward migration of the main channel is however limited by the underlying geology. The Pleistocene sea cliffs adjacent to the SR105 groin mark the furthest that the entrance channel to Willapa Bay has migrated to the north since the present high stand in sea level was reached several thousand years ago. If the channel should reach the base of these cliffs, they can be expected to exert increased, but not absolute, resistance to further northward migration.

In the shorter term, it does not seem likely that the main channel will migrate rapidly northwards in the next decade or so. This window of comparative stability would permit the construction of temporary flood protection works on Empire Spit that can adapt to change in shoreline position and configuration. Considering the possible recommencement of channel migration over longer timescales, it seems unwise to construct any more permanent flood protection works adjacent to the main channel, unless the works are designed to control the future migration of the main channel itself, an undertaking that is likely to be both extremely expensive and difficult to achieve.

Growth and erosion of Empire Spit

The future stability of Empire Spit is totally dependent on the future migration of the main channel. If the main channel recommences its northward migration, Empire Spit will erode rapidly and migrate parallel to the north edge of the channel. Any sand fill placed on top of Empire Spit will also be eroded and will not significantly affect the rate of northward migration of the channel as the quantity of sand transported in the channel is an order of magnitude greater than the quantity of sand that can be placed (or replaced) on the spit by mechanical means.

Replenishment of Empire Spit by a sand fill would (temporarily) help to limit wave penetration into North Cove during storm events. Whether the sand fill erodes at a rate faster or slower than the existing Empire Spit will depend to a large extent on the design of the fill. If the fill extends outward toward the main channel beyond the general alignment of the north edge of the main channel it can be expected to erode very rapidly. If, on the other hand, the fill is placed on top or 'inland' of the existing seaward limit of Empire Spit then it will tend to be more stable. The cross-sectional slope of the outer edge of the fill is another important design parameter. If this slope is steeper than the existing beach on Empire Spit then it may erode very rapidly during storm events by loosing sand across shore into the deep channel as the storm waves rapidly flatten the profile to a more stable angle.

4.6   Discussion

Application of the new Delft3D-FLOW model to the real-life erosion problems at Willapa Bay was a valuable exercise. Application of the model as part of a larger study and in conjunction with a coastal process measurement campaign contributed greatly to the understanding of the causes of and likely future for channel migration at the entrance to Willapa Bay. Application of the model also revealed the strengths and limitations of the present state of the art in coastal morphological modelling.

Obtaining coastal process measurements is complex, expensive, and fun. The potential to make errors in the collection and processing of measurement data is high and checking the measurements against good process models revealed several errors made during data processing. Having field measurements available against which to validate the numerical process models against added considerable confidence to the modelling, although they did not actually greatly affect the calibration of the underlying process models which were surprisingly good "out of the box".

Overall, the model was able to be successfully applied to the complex and dynamic environment of Willapa Bay. The model was relatively straightforward to set up and the underlying processes of tide and wave propagation were easy to calibrate against the available measurements. The basic process models were run at or near their default settings and reproduced the process measurements very well. Even observed residual currents were generally well reproduced by the combined wave and current models. Some complex interactions between fundamental process models were encountered which somewhat complicated the model calibration. Introduction of waves, and the associated enhanced bed roughness, upset the calibration of the tidal model for example. Prior documentation of expected process interactions and establishment of a recommended order of process model calibration would assist modellers seeking to calibrate complex process-based coastal morphological models.

Use of SWAN in a "quasi-nonstationary" manner appeared to work satisfactorily in the case of Willapa Bay. However the use of just one iteration in a stationary SWAN simulation is not a recommended manner of running SWAN and it is desirable that the linkage between Delft3D and SWAN be improved to allow the use of fully non-stationary SWAN simulations within the Delft3D framework.

The complete morphological model produced mixed results. It performed well (qualitatively) over the more recent period of 1998-2003 reproducing most of the observed behaviour. It did not perform so well starting from the less well known 1936 bathymetry and failed to reproduce the observed rapid northward migration of the entrance channel occurring at that time.

The behaviour of the main channels in the morphological model was poor using the default model settings. All channels tended to become too deep and narrow compared to observations. This was overcome by increasing down-slope bedload transport in the model. However, to achieve realistic morphological behaviour, model parameters had to be set to unrealistically high values. This indicates that a down-slope sediment transport process is probably missing from the model and is critical to the long-term morphological development of dynamic tidal channels.

In contrast to the calibration of the underlying process models, calibration of the overall morphological model was an extensive, painful, and ultimately not particularly successful exercise which did not contribute significantly to the study objectives. The process of calibrating individual processes within a morphodynamic feedback loop is still poorly understood and would benefit from systematic investigation and analysis. Some seemingly insignificant process calibration settings had a significant impact on morphological results (e.g. bottom roughness and wave streaming factors) however, despite extensive efforts, this study could not find the right "knobs" to adjust to successfully reproduce the observed 1936 morphological behaviour and this leads to the conclusion that important processes may be missing from the model and that others (e.g. wave enhancement of bed roughness) behave in an undesirable manner for a morphological model.

Despite the model's failure to reproduce the observed behaviour of the main channel, computed general patterns of sediment transport around the bay entrance are logical and robust and analysis of the differences in residual sediment transport pathways between 1936 and 1998 provides a convincing explanation of the observed channel behaviour.

Ultimately, application of the model to Willapa Bay achieved the objectives of the USGS study, however the required results were achieved by careful interpretation of the sediment transport pathways computed by the model, rather than by simply relying on the morphological model results themselves.

When applied to Willapa Bay, the morphological model results (as opposed to the residual sediment transport results) were most useful as a "reality check" on the accuracy and sensitivity of the computed sediment transport patterns, rather than as a direct morphological prediction method in and of themselves. However, having morphological model results, even poor ones, is definitely preferable to simply relying on the results of an initial sedimentation and erosion model.

It is likely that the poor performance of the morphological model is due to a missing process which critically affects the lateral migration and cross-sectional shape of tide-dominated channels. This is likely to be a fruitful area for further research and future model improvement.

The difficulty experienced calibrating the morphological model also casts doubt on the methods used to accelerate the morphological model (primarily wave input reduction and morphological acceleration). These techniques are investigated further in Chapter 5 to determine whether they are responsible for the poor performance of the morphological model.

Chapter 5

Morphological Acceleration Techniques

The use of process-based numerical morphological models to make both diagnostic and prognostic "brute-force" simulations of morphological evolution requires little explanation: the model is run for a designated simulation period starting from a specified initial condition. During the simulation time series of forcing conditions must be specified. For diagnostic simulations these forcing conditions normally consist of observed water-level, wind speed, atmospheric pressure, river discharge, and ocean swell conditions. For prognostic simulations, the stochastic nature of wind, wave and other forcing processes requires more careful consideration, often resulting in the use of forcing time series derived from mean annual or seasonal climates, and sensitivity testing of deviations around this mean.

Brute-force simulations such as these are useful for investigating the details of the time-histories of morphological developments and investigating the relative impact of the associated processes. However, they are extremely computationally intensive and will, depending on the resolution, range of processes simulated, and computer hardware available, often run at a speed in the order of just 10 times faster than real time. This means that when conducting studies of morphological evolution spanning time periods in excess of just a few weeks, techniques for accelerating the morphological evolution predicted by these models, or a great deal of patience, are required.

This chapter proposes improvements and modifications to a number of existing input reduction and morphological acceleration techniques in order to reduce the computational effort required to carry out medium-term morphological simulations with the new "online" sediment transport and morphological change model described in Chapter 2. The proposed techniques are applied one by one to a benchmark brute-force simulation and the resulting errors are analysed. The final result is an accelerated simulation which runs in less than 10% of the time of the brute-force simulation. This accelerated simulation is then used to test a number of further refinements to the proposed acceleration techniques.

5.1 Approaches and Considerations

A fundamental problem with performing process-based morphological modelling is that morphological evolution of coastal features of interest to engineers, managers, and the public at large usually occurs at time scales several orders of magnitude larger than the time scale of the hydrodynamic fluctuations driving the sediment transport. However, as discussed by de Vriend et al. (1993), this potential separation of scales also provides a basis for a range of morphological acceleration approaches. de Vriend et al. discuss three distinct approaches to accelerating long-term morphodynamic modelling:

1. Input reduction, which is based on the idea that the residual (long-term) effects of smaller-scale processes can be obtained by applying models of those smaller-scale process forced with reduced "representative" inputs.

2. Model reduction, which is based on the idea that the model itself can be reformulated at the scale of interest without describing the details of the smaller-scale processes.

3. Behaviour-oriented modelling, which attempts to model the phenomenon of interest without attempting to understand or describe the underlying processes.

In this chapter improved approaches to both input reduction and model reduction are investigated. Behaviour-oriented modelling is not discussed in this thesis, although the dry cell erosion technique implemented in the sediment transport model and discussed on page 27 could be viewed as a behaviour-oriented model component.

When attempting to reduce the complexity of a system of physical processes, it is essential to keep the scales of interest in mind at all stages of model selection and establishment. For example, it is clear that if one is interested in modelling the detail of the fluctuating morphology of seabed sand ripples under the action of a sequence of waves or wave groups it would be inappropriate to use a phase-averaged wave model. Equally, if one is interested in the episodic change in coastal morphology associated with the impact of a particular storm or spring tide, one should not attempt to do this using an averaged wave climate or representative morphological tide. de Vriend et al. were very clear on this point, stating

> "Reduction of models and data does not make sense unless it is intrinsically coupled to the formulation of an objective. This means we have to specify beforehand which aspects of the coastal behaviour we wish to focus on, and which we consider as unimportant."

This essential link between the objectives of a study and the applicability of acceleration methods is apparently missed by some authors. Winter et al. (2006) for example, who appear to decide a priori to use a "standard" method of input or model reduction and then attempt to evaluate the manners in which it fails to meet objectives that were not properly considered when the input reduction technique was selected.

The approach to morphological acceleration developed and utilised in Chapter 2 introduces a model reduction technique which uses a morphological acceleration factor or "morfac" to increase the rate of bed elevation change for each hydrodynamic time step, as more thoroughly described in Section 5.2.1 below. The use of morfac can be viewed as a form of model reduction as, conceptually, it decouples the time-scale of the morphological model from the time-scale of the hydrodynamic and sediment transport models. Once this decoupling occurs, it is important that the modeller realises that the most important time-scale of the model is now the morphodynamic time-scale and that the hydrodynamic and sediment transport models are only run in order to provide updates (in terms of residual sediment transport vectors) to the morphological model.

In order to effectively use this model reduction technique in a coastal setting it is necessary to also apply input reduction techniques to all forcing processes that are applied as boundary conditions to the wave, hydrodynamic and sediment transport models.

5.1.1   Previous approaches

Several approaches to model reduction are discussed by de Vriend et al (1993), Cayocca (2001), and Roelvink (2006). These approaches include

1. Increasing the effective time step of the morphological model by extrapolating bottom changes computed over one tidal cycle to several tidal cycles before recomputing the hydrodynamic, wave, and sediment transport fields. This can be achieved either by simple extrapolation, "elongating the tide" by progressively adding corresponding phases of multiple tidal cycles, or by various types of predictor/corrector approach.

2. Reducing the computational effort required to create the updated flow, wave, and sediment transport fields following changes in bottom elevation. These methods include the so-called "continuity correction" which can be used to approximate updates to hydrodynamic flow fields by assuming that $\vec{U}h$ is invariant at all points in the model domain, and the Rapid Assessment of Morphology (RAM) approach which directly updates sediment transport fields under the assumption that $\left|\vec{S}\right| = Ah^{-b}$ where A is a spatially-variable factor and b is a constant, both of which can be determined by initial hydrodynamic, wave, and sediment transport computations.

These approaches are frequently used in combination. For example, Cayocca (2001) applies the "elongated tide" method in combination with the continuity correction for modelling the morphological development of the Arcachon Lagoon over several years.

Use of any of these model reduction approaches also requires input reduction. Tidal forcing must be reduced to a morphologically representative tide, or tides, and wave forcing must be reduced to a limited set of representative wave conditions. In either case the objective is to select a limited subset of tides and waves which result in the same residual sediment transport (or morphological change) patterns as the complete tide and wave records over the period of interest.

A commonly adopted method for selecting a "morphological tide", as discussed by Latteux (1995), Cayocca (2001), and Grunnet et al. (2004), amongst others, is to simulate a lengthy period of the complete tidal record (a neap-spring cycle, for example) and then select the individual tidal cycle from this simulation which has a mean sediment transport pattern closest to the mean transport pattern of the full simulation. An adjustment factor may need to be applied to the resulting morphological time scale if the tidal cycle which most closely matches the mean sediment transport *pattern* does not also match the mean sediment transport *rate*. Following this approach, representative tides are commonly found to be somewhat (7 to 20%) larger than the mean tide (Latteux, 1995).

An alternative approach, the so-called "ensemble technique" used by Bernades et al. (2006), divides the full tidal signal into a number of discrete classes by tidal range. Individual tides falling within each class are then averaged, producing a "mean tide" for each tidal range class. The mean tides are then simulated consecutively, in ascending order of tidal range, with sediment transport and morphological change progressively updated during the simulation. Depending on the frequency of occurrence of each class of tidal range, mean tides may be repeated multiple times. A scaling factor, conceptually similar to morfac, may also be applied to each mean tide to reduce the number of mean tides that need to be simulated. Bernades et al. report good results when applying this method to 6-month simulations of the Teign estuary. Methods for combining the ensemble technique for tidal input reduction with wave input reduction were not discussed.

Selection of representative wave conditions for a morphological simulation has a similar objective to selecting a representative tide: to select a limited set of wave conditions, and associated weight factors, which will produce a similar mean sediment transport pattern to the full wave "climate" over the period of interest. Steijn (1992) suggests a method of selecting wave direction classes, representative wave heights, and weight factors which preserve both longshore power (SPM, 1984) and a wave stirring parameter (the wave-related bed shear stress, for example). More recently, Grunnet et al. (2004) used a separate beach profile model to compute longshore sediment transport for each of 77 wave direction and height classes. Four directional sectors were then determined and six representative wave conditions selected. The wave conditions were weighted in such a way that the longshore transport for each directional sector, and the wave climate as a whole, was preserved.

Approaches such as these have frequently been applied to "engineering" studies of coastal morphological problems. Roelvink et al. (1994) model a migrating tidal inlet after a breach; Steijn et al. (1998) investigate the impact of long dam near a Dutch tidal inlet. Gelfenbaum et al. (2003) model the effect of jetties on an inlet on the US west coast, and Cañizares et al. (2003) model morphological changes at Shinnecock Inlet, New York, USA. A similar approach is followed for the 5-year morphological simulations of Willapa Bay presented in Chapter 4 of this thesis. Results of these studies tend to be fair at best, with the morphological models usually reproducing many of the broad trends observed, but failing to reproduce many of the details. One of the open questions regarding such studies is whether the lack of model skill should be attributed to the input and model reduction techniques required to simulate coastal morphology for useful time scales, or whether it is due to shortcomings in the physical process descriptions contained within the numerical models themselves.

5.1.2 The present approach

Lesser et al., 2004 (and Chapter 2 of this thesis) present a new process-based morphological model which computes sediment transport and morphological change simultaneously, or "online", with the hydrodynamic processes. This approach has several advantages including simplicity of model setup and operation, and simulation of the full three-dimensional dynamic interaction of sediment and hydrodynamics (e.g. sediment density effects) on hydrodynamic time-scales. The online computation of morphological change also opened the way for the implementation of a new form of morphological acceleration by way of the so-called "morphological acceleration factor" or "morfac" introduced by Lesser et al. (2004). This approach is conceptually very similar to the elongated tide method proposed by Latteux (1995) however it has the advantage of not relying on use of the continuity correction to approximate updates to the flow field during tidal cycles. Early validation tests using this technique reported by Lesser et al. and Chapter 3 of this thesis include modelling the migration of a trench in a laboratory-scale flume (Section 3.3.1), the deformation of a Gaussian hump in a flume (Section 3.4.1), and the morphological changes around the breakwaters at the Dutch harbour of IJmuiden (Section 3.3.4). The tests demonstrated that the online morfac approach does indeed have the beneficial properties Latteux identified in the elongated tide approach: stability, little diffusion, and accurate propagation of bed forms.

The online morfac approach to morphological updating has since been used and reported by a number of authors: Reniers et al. (2004) applied the approach in conjunction with an advanced wave-group model to model the development of detailed near-shore morphology. Grunnet et al. (2004, 2005) applied the online morfac approach to simulating the morphological development of a shoreface nourishment and investigation of the relative importance of the various forcing processes. Recently, van der Wegen and Roelvink (2008) applied the morfac approach to modelling several centuries of evolution of a schematised tidal basin. There has, however, only been rather limited error analysis of the use of the online morfac approach. Roelvink (2006) compared a number of morphological updating techniques, including the online morfac approach, performed a qualitative theoretical accuracy analysis, and compared the online morfac approach with an extension of the method – the so-called "parallel online" approach – applied to the morphological development of a single point in a highly schematised tidal inlet. There have not however been any quantitative analyses of the error introduced by the online morfac or parallel online approaches to morphological updating applied to real coastal environments with detailed wave climates and/or a comparison with the error introduced by other aspects of the morphological acceleration technique, such as the input reduction required to effectively use any of the morphological acceleration techniques. That is the central objective of this chapter.

5.1.3   Validation of the techniques

The test-bed used for this investigation of the errors introduced by morphological acceleration techniques is the Delft3D morphological model of Willapa Bay, WA, USA described extensively in Chapter 4. Willapa Bay is a large meso-tidal estuary the entrance of which is exposed to the vigorous wave climate of the north-east Pacific (see Figure 4.1 on page 57. The entrance of the bay is a naturally dynamic pattern of channels and shoals consisting of well-sorted fine sand. Little dredging is carried out in Willapa Bay, and none in the bay entrance. The bay has been the subject of several major field experiments, is surveyed regularly, and is the site of a long-term tide gauge. A directional wave buoy is also located nearby offshore. All this available data, combined with the dynamic nature of the bay entrance, make Willapa Bay an ideal subject for the development and testing of medium-term morphological models. Although the techniques discussed here are validated using this specific Delft3D model, they are generally applicable and could be applied to other locations and other modelling systems.

In order to isolate and quantify the impact of the various input reduction and morphological acceleration techniques required to accelerate the simulation of five years of morphological development of the entrance to Willapa Bay this study takes a novel, but straight-forward, approach. A brute-force simulation of the morphology of the entrance of Willapa Bay is made with a directly-coupled hydrodynamic, wave, sediment transport, and morphological model subject to the full recorded time series of forcing processes over the five year period from November 1998 to November 2003. Despite the model's shortcomings in terms of initial conditions and physical process descriptions, this simulation is then regarded as a perfect benchmark simulation for the purposes of testing input reduction and morphological acceleration techniques. Simplified time series of forcing processes are subsequently introduced one by one into the brute-force simulation and the results are compared to the benchmark simulation. Once all the simplified forcing processes have been introduced independently they are combined and simulated again to test for non-linear interactions between the individual input reductions. Finally, the full set of simplified forcing processes is combined with the variable morfac morphological acceleration technique and the results are compared with both the final and initial brute-force simulations. The former comparison isolates the error introduced by just the variable morfac acceleration technique, whereas the latter comparison allows quantification of the total error introduced by the complete package of input reduction and morphological acceleration techniques required to run an accelerated morphological simulation. The accelerated simulation requires approximately 10% of the time and computational effort of the brute-force simulations. A series of accelerated simulations is then performed to test possible improvements and sensitivity to the wave, wind, and tidal input reduction techniques employed.

Errors introduced by each stage of the input reduction and morphological updating techniques are quantitatively compared in order to identify which aspects of these techniques require the most careful consideration when constructing long-term simulations and to identify which techniques are most likely to reward further research effort.

5.1.4   Structure of this chapter

The methods of input and model reduction used to accelerate the 5-year morphological simulations of Willapa Bay are discussed in detail in Section 5.2 where the methods used to evaluate the error introduced into the simulation by each of these simplifications are also described. The results and an analysis of the differences between the various model simulations are presented in Section 5.3. The results of an investigation into the sensitivity and optimisation of the acceleration techniques are also presented in this section. The main conclusions of this chapter are discussed in Section 5.4.

5.2   Methods

5.2.1   Morphological acceleration

The "morfac" approach to morphological acceleration developed and discussed by Lesser et al. (2004) and in Chapter 2 of this thesis is attractive because it is numerically efficient, robust, and simple to implement. Conceptually the idea is very straightforward in simple situations with stationary boundary conditions, such as many of the validation tests discussed in Chapter 3. When applied to coastal situations subject to oscillating tidal currents and other time-varying forcing conditions the concept becomes slightly more complex and deserves some elaboration.

The basic concept

As described in Section 2.6.2, the morfac approach works by multiplying the suspended sediment erosion and deposition fluxes and the gradients in the bed-load sediment transport vector components by a spatially-constant factor (morfac). Doing this effectively multiplies any computed change in bed elevation occurring during a hydrodynamic time step by the factor morfac and can be viewed as effectively making the time step of the morphological model morfac times greater than the hydrodynamic time step. When fluctuating boundary conditions are applied to the hydrodynamic model it is essential to conceptually separate the hydrodynamic and morphological models and their associated time scales. Fluctuations in boundary conditions are applied on hydrodynamic time scales, however the impact of these fluctuations are amplified by morfac before being applied to the morphological model.

When applied in a tidal environment, the morfac approach to morphological acceleration results in a situation rather similar to the "lengthening of the tide" approach described by de Vriend et al. (1993), Latteux (1995) and Cayocca (2001). However, unlike the approach described by these authors, the morfac approach does not use a "continuity correction" to approximate the changes in flow velocities at intermediate (intra-tide) morphological time steps but, rather, it updates the full non-linear hydrodynamic simulation by taking the next hydrodynamic time step on the updated bathymetry. With this in mind, in an accelerated morphological simulation, the hydrodynamic model including intra-tide bathymetry updates, should be regarded as a "sub-grid" model for the morphological model which steps forward with time increments of morfac × tides. The morfac approach displays the same positive characteristics that Latteux identified in the "lengthening of the tide" approach:

1. It allows bed forms to propagate during the tidal cycle.

2. It is extremely stable as the very frequent adjustments to the hydrodynamics provides a natural feedback mechanism to damp out any instabilities in the bathymetry that may develop.

3. It creates little numerical diffusion.

The morfac approach also suffers from the same limitations as "lengthening of the tide", the principle of which is that an implicit *assumption of linearity over the lengthened tidal cycle* is required, in order to allow for the superposition of the various contributions to the bed level change in an arbitrary sequence (de Vriend et al., 1993). This requires that bed elevation changes and the changes to the associated sediment transport patterns must be able to be assumed to be approximately linear over the full sequence of morfac × tides. In the words of Roelvink (2006)

> *"the idea is that nothing irreversible happens within an ebb or flood phase, even when all changes are multiplied by [morfac]."*

This also means that bed elevations computed *during* tidal cycles are not realistic because only a subset of the full cycle of the arbitrary sequence of tidal phases have been added. They are relevant only for the stability and accuracy of the hydrodynamic and sediment transport "sub-grid" models and should be ignored completely when evaluating the performance of the morphological model.

The validity of this assumption of linearity can clearly be violated by choosing a morfac that is too high. The upper level for morfac, at which the system can no longer be assumed to behave in an approximately linear manner, will vary with the stability of the situation being modelled and will need to be assessed on a case by case basis. The validity of the assumption of linearity can easily be tested, however, by conducting repeated simulations with different morfac values and appropriately adjusted hydrodynamic simulations. The two simulations should produce virtually identical results if the system is behaving in a linear manner. A test such as this should be conducted early in any study where a morphological acceleration factor is used. As an indication, however, tests performed to date have indicated that morfac in the range of 10 to 100 can usually be safely applied in coastal zones where waves are significant. For example, see Lesser et al. (2003), Grunnet et al. (2004), Reniers et al. (2004), and Chapter 4 of this thesis. This effectively extends the useful morphological simulation duration for these environments to around a decade. Longer durations could be achieved using greater computational resources and the "parallel online" approach (an extension of the basic morfac approach) described by Roelvink (2006). In less morphologically dynamic simulations, where the effect of waves can be neglected for example, higher morfac values can be employed. Van der Wegen and Roelvink (2008) validate and use morfac values of up to 400, allowing them to simulate up to 8,000 years of estuarine profile development in a one-dimensional model and up to 1,600 years in two dimensions. Similar results have been achieved by Dastgheib et al. (2008) who used the "online" approach and a morfac of 300 to simulate 2,100 years of morphological development of tidal basins in the Dutch Wadden Sea subject to simple tidal forcing.

Variable morfac, a word of caution

For some applications it is desirable to change the morfac value during the course of a morphological simulation. This is possible, and simply makes the ratio between the morphodynamic and hydrodynamic time steps a function of (morphological) time. As long as a robust method of keeping track of morphological time is implemented, as is the case in Delft3D, changing morfac is conceptually not complicated. There is one caveat however. When morfac changes and sediment is in suspension continuity of sediment mass is not conserved. This is easily demonstrated by a simple example: Imagine a closed system, such as an annular flume. If a simulation with initially clear water and 100 kg/m^2 of sediment available on the bed is accelerated from rest with a morfac of 10 until uniform depth-integrated suspended sediment loads reach 2 kg/m^2 the quantity of sediment remaining at the bed would be $100 - (2 \times 10) = 80$ kg/m^2. If morfac was then changed to 100 and the flow allowed to come to rest and the sediment settle (uniformly) back down to the bed the final quantity of sediment at the bed would be $80 + (2 \times 100) = 280$ kg/m^2. Clearly sediment mass has not been conserved. This occurs because sediment in suspension (including suspended sediment entering or leaving the model through lateral boundaries or internal sources) effectively *represents* morfac $\times$ its own mass. Thus, if morfac is changed while a depth-integrated mass of sediment S per m^2 of bed is in suspension, a sediment mass error equal to $S \times \Delta$morfac per m^2 of bed will occur. It is important to note that this problem also applies to the implicit "change" from and to morfac = 0 at the start and end of every morphological simulation.

The discontinuity in sediment mass in the simulation could be solved by modifying the sediment bookkeeping scheme to add $(1-\text{morfac}) \times S$ to the bed sediments in each grid cell. This would prevent the loss of continuity, but would not solve the fundamental problem and would cause sudden changes in bed elevation to occur at times when morfac was changed. Complications would also occur over areas with limited sediment availability at the bed. On the balance, it was decided to leave the possibility of a sediment mass error occurring at times of changing morfac in Delft3D. The problem can be minimised by carefully choosing the start and end times of a morfac value so that suspended sediment concentrations are a) relatively low and b) approximately equal. The design of the accelerated simulations discussed in Section 5.2.6 gives an example of how this can be achieved.

Application to coastal situations

In order to use a morphological acceleration technique in a coastal situation it is essential to identify which coastal processes play a significant role in (residual) sediment transport patterns over the space and time scales of interest. Existing literature may already identify the processes of relevance, otherwise, a series of brute-force sediment transport simulations subject to all forcing processes may be performed in order to test the (spatial) impact of eliminating individual forcing processes. In most coastal situations sediment transport will be governed by tidal currents, often with the addition of waves, wind, and/or river discharge. In more complex environments three-dimensional density currents, stratification, and internal waves may also play an important role.

The next step is to judge what spatial and temporal resolution of morphological developments needs to be resolved in order to predict or reproduce the morphological behaviour of interest. For example, is the morphological development dominated by episodic events or can the effects of individual events be averaged together and modelled as gradual change? This step is making an assessment of the linearity of the morphological development. This is an exact parallel to the discussion regarding the assumption of linearity over a morphological tide above: over what scales can the chronology of the morphological developments be ignored as long as the individual contributions are modelled accurately and superimposed? Working upward through the relevant process scales may be a helpful approach: Is the sequence of individual waves or wave groups likely to be important (and the associated morphological change "irreversible")? The sequence of tides within the neap-spring tidal cycle? Change occurring during individual storm events? During individual seasons? Individual years? Often existing literature regarding the morphological feature(s) of interest will give a good indication of the important scales of forcing. Any judgements made at this stage can be checked by using the model. For example, if it is felt that seasonality is not an important factor and that a simple mean (wave) climate will be used, then a simulation using a chronology of wave climates for the individual seasons can also be performed. Differences observed can then be judged for significance relative to the scale of the features of interest. The results of tests similar to this applied to Willapa Bay are discussed in Section 5.3.8.

The goal of the above analysis is to identify a) which forcing processes need to be included in the morphological model and b) to what extent the complexity of the chronology of each of the important forcing processes can be reduced to "representative" forcing conditions which ignore the chronology while conserving the long-term residual sediment transport rates and patterns. If *all* forcing processes can either be neglected or simplified at a particular temporal scale then this opens the way to the application of the input reduction and morphological acceleration techniques discussed in the following sections.

5.2.2   Tidal input reduction

The objective of tidal input reduction is to replace the complex time series of tidal water level and current fluctuations occurring in nature with a simplified tide or tides. Ideally, the simplified tide should produce the same residual sediment transport and morphological change patterns as the naturally varying tides at all points over the region of interest and for the time period in question. If other forcing processes (e.g. waves) are significant then the randomness of the relative phasing of the tide and other forcing process will also need to be considered. If any systematic phase relationship is not automatically accounted for by processes represented in the model it will need to be explicitly specified in the forcing conditions. For example, care must be taken when strong diurnal sea breezes occur and interact in a consistent manner with a significant diurnal component in the tide.

Many authors have used tidal input reduction to extend the horizon of their morphological models into the medium term. The technique most commonly described in the literature (e.g. Gelfenbaum, 2003 and Grunnet, 2004) is based on the idea of Latteux (1995) that a morphological tide can be chosen where the pattern of sediment transport, or morphological change, over the area of interest most closely matches the pattern of transport, or morphological change, over an entire neap-spring tidal cycle. Most authors select just a single morphological tide, even though Latteux found that in situations with complex bathymetry a single representative tide may be insufficient to accurately represent sediment transport patterns over the complete domain. Under these situations Latteux recommended using at least two representative tides, one around mean or neap tidal range, and the other close to spring. Once the tide, or tides, which most closely match the *pattern* of the morphological change occurring over the full neap-spring tidal cycle are found, a scale factor is also identified to account for any discrepancy in morphological development *rate*.

An improved representative tide

A morphological tide resulting from the "standard" method for tidal reduction described above was initially applied to the Willapa Bay model, as described in Chapter 4. However, when the result of a brute force simulation containing this schematised tide was compared to the benchmark simulation, it was clear that important residual sediment transports were not accurately captured by the morphological tide. Investigation of this problem revealed that the error in residual transport occurred only when certain components were omitted from the full set of tidal harmonics used to drive the hydrodynamic model. Specifically, the missing residual was caused by missing the interaction of the M_2, O_1, and K_1 tidal constituents. This is logical, as the sum of the frequencies of the O_1 and K_1 constituents is exactly equal to the frequency of the M_2 constituent, which means that the interaction of these constituents does not cancel out in the long term.

The interaction of the M_2, K_1, and O_1 constituents, and implications for residual sediment transport, have been previously identified and discussed by Hoitink et al. (2003) who explain that the cause of the phenomenon is related to the fact that the M_2, K_1, and O_1 constituents are composed of simple linear combinations of just two of the six fundamental astronomical frequencies (see Pugh, 1987) and can therefore combine in a manner which produces a regular asymmetric neap-spring cycle with a period of 13.66 days. They point out that the residual sediment transport caused by this non-linear interaction can be expected to be more important than the well-known residual transport due to the non-linear interaction of the M_2 tide with the M_4 overtide if $2O_1K_1 > M_2M_4$ where O_1, K_1, M_2, and M_4 are the amplitudes of each of the respective tidal constituents. This will be the case in most locations on the west coast of the continental United States, as well as in many other locations around the world where the diurnal tidal constituents O_1 and K_1 are significant. Hoitink et al. go on to identify other tidal constituents which can interact in a similar manner. The amplitudes of these other constituents are generally very minor relative to the main M_2, K_1, and O_1 constituents and the residual effects caused by these other interactions are not considered further in the present study.

Hoitink et al. performed their analysis by analytically deriving expressions for the long-term average of U^3 and U^5 as the rate of sediment transport can generally be assumed to be proportional to flow velocity raised to a power in this vicinity. They do not, however, suggest a method for representing this extra residual transport in a simplified tide, which is what is required for the purposes of tidal input reduction.

To see if the residual sediment transport caused by the interaction of O_1, K_1, and M_2 can be captured in a simplified term the present analysis also assumes that sediment transport is proportional to U^3. If the velocity components are given by simple harmonic terms, then the residual sediment transport would be given by

$$\left\langle [O_1 \cos(\omega_{O1} t - \phi_{O1}) + K_1 \cos(\omega_{K1} t - \phi_{K1}) + M_2 \cos(\omega_{M2} t - \phi_{M2})]^3 \right\rangle \quad (5.1)$$

where $\omega_{O1}, \omega_{K1}, \omega_{M2}$ are the angular frequencies, O_1, K_1, M_2 are the amplitudes, and $\phi_{O1}, \phi_{K1}, \phi_{M2}$ are the phase offsets of the harmonic velocity constituents. $\langle \rangle$ denotes taking a long-term average.

Expanding this expression, keeping the O_1 and K_1 terms together, results in

$$\left\langle [O_1 \cos(\omega_{O1} t - \phi_{O1}) + K_1 \cos(\omega_{K1} t - \phi_{K1})]^3 \right\rangle +$$
$$+ \left\langle 3 [O_1 \cos(\omega_{O1} t - \phi_{O1}) + K_1 \cos(\omega_{K1} t - \phi_{K1})]^2 [M_2 \cos(\omega_{M2} t - \phi_{M2})] \right\rangle +$$
$$+ \left\langle 3 [O_1 \cos(\omega_{O1} t - \phi_{O1}) + K_1 \cos(\omega_{K1} t - \phi_{K1})] [M_2 \cos(\omega_{M2} t - \phi_{M2})]^2 \right\rangle +$$
$$+ \left\langle [M_2 \cos(\omega_{M2} t - \phi_{M2})]^3 \right\rangle$$
$$(5.2)$$

in which the first, third, and fourth terms have zero time-average. Expanding the remaining (second) term produces

$$3 \left\langle \left[O_1^2 \cos^2(\omega_{O1} t - \phi_{O1}) + 2 O_1 K_1 \cos(\omega_{O1} t - \phi_{O1}) \cos(\omega_{K1} t - \phi_{K1}) + \right. \right.$$
$$\left. \left. + K_1^2 \cos^2(\omega_{K1} t - \phi_{K1}) \right] [M_2 \cos(\omega_{M2} t - \phi_{M2})] \right\rangle \quad (5.3)$$

in which the first and third elements of the first term also have zero time-average when combined with the M_2 term. Leaving the following expression

$$3 \left\langle \left[O_1 K_1 \cos((\omega_{O1} + \omega_{K1}) t - (\phi_{O1} + \phi_{K1})) + \right. \right.$$
$$\left. \left. + O_1 K_1 \cos((\omega_{O1} - \omega_{K1}) t - (\phi_{O1} - \phi_{K1})) \right] [M_2 \cos(\omega_{M2} t - \phi_{M2})] \right\rangle \quad (5.4)$$

Recognising that $\omega_{O1} + \omega_{K1} = \omega_{M2}$ allows us to rewrite as

$$3 O_1 K_1 M_2 \left\langle \cos(\omega_{M2} t - (\phi_{O1} + \phi_{K1})) \cos(\omega_{M2} t - \phi_{M2}) \right\rangle \quad (5.5)$$

by applying basic trigonometry and recognising that $\langle \sin(\omega_{M2} t) \cos(\omega_{M2} t) \rangle = 0$

this can be simplified to

$$3O_1K_1M_2 \left[\left\langle \cos^2\left(\omega_{M2}t\right)\cos\left(\phi_{O1}+\phi_{K1}\right)\cos\left(\phi_{M2}\right) \right\rangle + \right.$$
$$\left. + \left\langle \sin^2\left(\omega_{M2}t\right)\sin\left(\phi_{O1}+\phi_{K1}\right)\sin\left(\phi_{M2}\right) \right\rangle \right] \tag{5.6}$$

in which $\left\langle \cos^2(\omega_{M2}t) \right\rangle$ and $\left\langle \sin^2(\omega_{M2}t) \right\rangle$ both equal $\frac{1}{2}$. Leaving

$$\frac{3}{2}O_1K_1M_2 \left[\cos\left(\phi_{O1}+\phi_{K1}\right)\cos\left(\phi_{M2}\right) + \sin\left(\phi_{O1}+\phi_{K1}\right)\sin\left(\phi_{M2}\right)\right] \tag{5.7}$$

which, using basic trigonometry, can be simplified to

$$\boxed{\left\langle U^3_{O1K1M2} \right\rangle = \frac{3}{2}O_1K_1M_2\cos\left(\phi_{O1}+\phi_{K1}-\phi_{M2}\right)} \tag{5.8}$$

which implies that the magnitude and sign of the residual sediment transport due to the interaction of the O_1, K_1, and M_2 tides are dependant on the magnitude and relative phases of these three constituents – as previously pointed out by Hoitink et al.

The present approach then attempts to capture the residual sediment transport given by Equation 5.8 by replacing the O_1 and K_1 components with an artificial diurnal tidal constituent C_1 which has an angular frequency $\omega_{C1}=0.5\omega_{M2}$, giving the following expression for the velocity raised to the third power

$$\left\langle \left[C_1\cos\left(\omega_{C1}t-\phi_{C1}\right) + M_2\cos\left(\omega_{M2}t-\phi_{M2}\right)\right]^3 \right\rangle \tag{5.9}$$

Simplifying this expansion is similar to that of the M_2/M_4 coupling which has previously been described by van de Kreeke and Robaczewska (1993). Essentially, it expands to

$$\left\langle C_1^3\cos^3\left(\omega_{C1}t-\phi_{C1}\right) \right\rangle +$$
$$+ \left\langle 3C_1^2\cos^2\left(\omega_{C1}t-\phi_{C1}\right)M_2\cos\left(\omega_{M2}t-\phi_{M2}\right) \right\rangle +$$
$$+ \left\langle 3C_1\cos\left(\omega_{C1}t-\phi_{C1}\right)M_2\cos^2\left(\omega_{M2}t-\phi_{M2}\right) \right\rangle + \tag{5.10}$$
$$+ \left\langle M_2\cos^3\left(\omega_{M2}t-\phi_{M2}\right) \right\rangle$$

in which the first, third, and forth terms have zero long-term average and vanish. Simplifying the remaining, second, term results in

$$\boxed{\left\langle U^3_{C1M2} \right\rangle = \frac{3}{4}C_1^2M_2\cos\left(2\phi_{C1}-\phi_{M2}\right)} \tag{5.11}$$

Equating the residuals given by Equations 5.8 and 5.11

$$\frac{3}{2}O_1K_1M_2\cos\left(\phi_{O1}+\phi_{K1}-\phi_{M2}\right) \quad = \quad \frac{3}{4}C_1^2M_2\cos\left(2\phi_{C1}-\phi_{M2}\right) \tag{5.12}$$

requires that

$$\boxed{C_1 = \sqrt{2O_1K_1} \quad \text{and} \quad \phi_{C1} = \frac{\phi_{O1}+\phi_{K1}}{2}} \tag{5.13}$$

where C_1 and ϕ_{C1} are the amplitude and phase of a diurnal constituent with a tidal period of $2 \times M_2$ which will interact with M_2 to produce the same third order velocity moment, and therefore similar residual sediment transport, as the O_1 and K_1 tidal constituents.

This result implies that a representative tide, which includes the residual third order moment due to the O_1, K_1, and M_2 interaction term, can be achieved by applying a simple repeating *double* tide consisting of only the M_2 and C_1 constituents. This tide will have a period of 24 hours 50 minutes 28 seconds (1490.47 minutes) and will display a daily inequality, the magnitude of which will depend on the relative amplitude and phasing of the M_2, O_1, and K_1 tidal constituents.

Complication – interaction with mean flows

A mean flow (non-tidal residual) causes enhanced residual transports in the presence of harmonic tidal velocity fluctuations due to the non-linear nature of sediment transport. Assuming sediment transport is proportional to U^3 an additional residual term is created which is proportional to $U \times (M_2^2 + S_2^2 + N_2^2 + O_1^2 + K_1^2 + \ldots)$. This can be seen from the expansion

$$\langle [U + a]^3 \rangle = U^3 + 3U^2 \langle [a] \rangle + 3U \langle [a]^2 \rangle + \langle [a]^3 \rangle \tag{5.14}$$

where $a = [M_2 \cos(\omega_{M2} t + \phi_{M2}) + S_2 \cos(\omega_{S2} t + \phi_{S2}) + \ldots]$.

The first term in this expansion will be accurately captured by the model as long as the residual current is accurately reproduced. The second term is zero due to averaging, the third term survives, and the forth term contains the tidal asymmetry including the O_1, K_1, and M_2 interaction discussed above. Thus, for a simplified tide to create the same residual sediment transport as a full tide in the presence of a non-tidal residual it is important that the sum of the squares of the constituent amplitudes (the tidal energy) is preserved in the simplified tide.

Clearly, a simplified tide of $M_2 + C_1$ alone does not satisfy this requirement. Matters can be improved however by applying a scaling factor to the M_2 and/or C_1 constituents to preserve total energy. This is not a perfect solution however, as applying the factor to bring energy levels up to the required level will overstate the residual due to $M_2 + O_1 + K_1$ interaction. If applied to both M_2 and C_1 constituents the required factor is

$$f_1 = \sqrt{\frac{(M_2^2 + S_2^2 + N_2^2 + O_1^2 + K_1^2 + \ldots)}{M_2^2 + C_1^2}} \tag{5.15}$$

alternatively, if the factor is to be applied only to the M_2 constituent it will need to satisfy

$$f_2 = \sqrt{\frac{(M_2^2 + S_2^2 + N_2^2 + O_1^2 + K_1^2 + \ldots) - C_1^2}{M_2^2}} \tag{5.16}$$

To check the error introduced by either of these approaches, if a scaling factor f_1 is applied to $(M_2 + C_1)$ and $f_1 M_2$ and $f_1 C_1$ are substituted for M_2 and C_1

in Equation 5.9 then the tidal residual given by Equation 5.11 will be increased in proportion to f_1^3. If, on the other hand, the slightly larger scaling factor f_2 is applied to only the M_2 component then the tidal residual will only be increased linearly with f_2. This latter option can therefore be expected to introduce less error into regions of simulations with little residual flow.

To summarise:

1. To conserve residual transports using a simplified tide in areas with significant residual currents the total energy of the full tide (the sum of the squares of the amplitudes of the tidal constituents) should be conserved in the simplified tide. This could be achieved by applying an amplification factor to the M_2 and/or C_1 tidal constituents.

2. Applying the factor to just the M_2 component means that the error introduced into the M_2–O_1–K_1 residual will be linear with the amplification factor. This is preferable to applying the factor to both the M_2 and C_1 constituents which makes the error proportional to the amplification factor cubed.

3. The factor to be applied is a function of the mean flow. if $U = 0$ then $f_1 = f_2 = 1.0$. if U is large then f_2 should approach the limiting value given by Equation 5.16.

4. An analytical expression *could* be derived which determines the "correct" f_2 factor for a given set of harmonics and mean flow for any given point in a model. However the expression would be complicated and the "correct" f_2 value would vary spatially, depending on the relative strength of the mean flow.

5. The "correct" f_2 value for any given location would not in fact be completely correct as the derivations above are based on the assumption that sediment transport is proportional to U^3. This is not strictly correct and will vary depending on the dominant mode of sediment transport (suspended vs bed load) and factors such as the intensity of wave stirring.

6. A pragmatic method to determine an optimum amplification factor is therefore required. Trial and error comparison of a range of factors against a brute-force simulation could be a straightforward and practical solution.

A general method to determine a morphological tide

Given this new insight into the residual sediment transport caused by tidal harmonics in mixed tidal regimes, the recommended procedure for selecting a morphological tide is to calculate the required M_2 and C_1 harmonics using Equation 5.13. Over-tides of M_2 should also be included if they are significant. The optimum factor f_2 should then be determined through trial and error. A brute-force simulation including typical levels of other important forcing processes, such as waves, should be simulated for at least the duration of a neap-spring cycle. The simulation should be repeated using morphological tides consisting of $f_2 \times M_2 + C_1 + M_2$ overtides using a range of f_2 factors. These simulations could be accelerated to save computational effort. The optimum factor will typically be close to that given by Equation 5.16 if non-tidal residuals are significant, and closer to 1.0 if they are weak. Choose the factor which most closely matches the pattern of the neap-spring bathymetry changes (e.g. minimises MSE). The gradient of the correlation line should be used as an additional time-scale factor if it significantly deviates from 1.0.

An improved morphological tide for Willapa Bay

The WBAY model of Willapa Bay (described in Chapter 4) on which the morphological model used in this chapter is based, is driven by an offshore water level boundary condition prescribing water levels which are a linear interpolation between the water levels prescribed at the north-western and south-western corners of the model. The water levels at these corners were obtained from the larger scale ORWA2 model, also described in Chapter 4. To construct a morphological tide using the new method, the ORWA2 model was run for a period of 89 days and the water levels at the locations of the corner points of the WBAY model were analysed using a harmonic analysis program to obtain astronomical tidal constituents at these points (Table 5.1). Assumed amplitude ratios and relative phases, based on the nearby Toke Point tide station, were used to infer the amplitudes and phases of the P_1, K_2, and NU_2 constituents based on the K_1, S_2, and N_2 constituents respectively. The amplitude and phase of the artificial C_1 constituent were then calculated using Equation 5.13, and are also shown in Table 5.1. The frequencies of the M_2 and C_1 components were adjusted slightly to make the morphological tide periodic at exactly 1490 minutes, with the adjusted M_2 component having a period of exactly 745 minutes. As a reference brute force simulation was available for the Willapa Bay model, the scale factor for the morphological tide was determined by performing a series of simulations using $1.06 \times (M_2 + C_1)$, $1.08 \times (M_2 + C_1)$, and $1.10 \times (M_2 + C_1)$. Comparison of the residual sediment transport through the control sections shown in Figure 5.1 (results not repeated here) confirmed the choice of 1.08 as the optimum choice for amplification factor f_1 – slightly larger than the theoretical value of 1.07 calculated using Equation 5.15. The morphological tide of $1.08 \times (M_2 + C_1)$ is used in brute force simulations $bf6$ and $bf7$ and most of the accelerated simulations discussed in this chapter. The alternative use of a morphological tide using the f_2 factor is tested using accelerated simulations and is discussed in Section 5.3.9. The full natural tidal signal and simplified morphological tide over a period of 17 days are illustrated in Figure 5.2.

5.2.3   Wave input reduction

The objective of wave input reduction is to define a limited number of offshore wave classes which together produce the same residual sediment transport patterns and rates as the full time series of offshore wave conditions over the time period of interest. Unlike tide, the chronology of future wave conditions is not known although a statistical description of the prevailing wave "climate" can usually

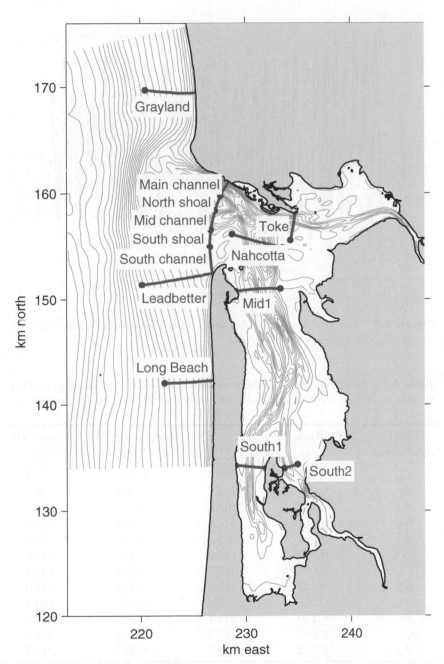

Figure 5.1 – *Overview of Willapa Bay showing sediment transport control section locations.*

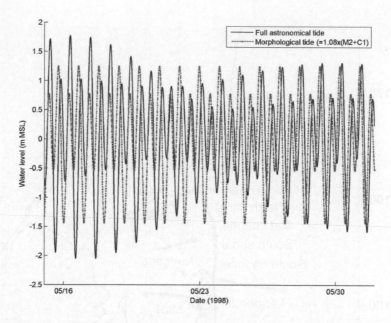

Figure 5.2 – *Full astronomical and simplified morphological tide shown for the duration of one neap-spring tidal cycle.*

Table 5.1 – *Tidal constituents applied at north-west (NW) and south-west (SW) corners of WBAY model. Analysed from an 89 day ORWA2 model simulation.*

Tidal Constituent	Amplitude SW (m)	Phase SW (°)	Amplitude NW (m)	Phase NW (°)
M_2	0.908	229.3	0.930	230.4
S_2	0.256	254.7	0.263	256.1
N_2	0.180	203.8	0.183	205.0
K_2	0.067	247.6	0.069	249.0
NU_2	0.041	204.3	0.042	205.5
$2N_2$	0.032	197.3	0.031	197.6
L_2	0.026	237.6	0.026	239.5
K_1	0.431	237.6	0.433	237.0
O_1	0.258	227.2	0.267	229.3
P_1	0.130	234.1	0.131	233.5
Q_1	0.050	227.4	0.049	228.0
J_1	0.032	244.0	0.032	243.5
C_1	0.472	232.4	0.481	233.2

be obtained from analysis of historical wave measurements. Thus, two problems present themselves when attempting to schematise a wave climate:

1. At what scale does the chronology of wave events become important? and

2. At scales in which the chronology of wave events can be assumed to be unimportant, what representative set of wave classes can be used to produce the same residual sediment transport and morphological change patterns over the area of interest as the full wave climate?

An additional complication may arise if wave conditions at the model boundaries vary on similar time scales to and have a non-random phasing with other forcing processes such as wind, tide, or river discharge. If such a non-random phase relationship occurs it will also need to be captured in the reduced input conditions.

The objective of this study was to model the morphology of the entrance to Willapa Bay, with a length scale of several kilometres, for the duration of five years. The morphological changes of interest are the migration of main channels and the ebb and flood tidal shoals. Previous literature (Ruggiero et al., 2005) had identified a strong seasonal (winter/summer) cycle in forcing processes with low frequency tidal levels, wind, wave height, direction, and period, and river discharge all showing strong seasonal cycles. Major changes, or "switching", of outer channel alignment had also been correlated with fluctuations of wave climate on inter-annual scales (Hands and Shepsis, 1999). van Overeem et al. (1992) have also shown that capturing the intermittence of storms and calmer periods is essential for modelling sediment transport around the outer delta of a tidal inlet in the Dutch Wadden Sea. Therefore, it is reasonable to expect that capturing the (mean) fluctuation in forcing processes between winter and summer seasons might well be important for modelling several years of morphological development. For diagnostic simulations of the historical period from 1998 to 2003 a strong inter-annual signal was also apparent, as 1998 was part of an El Niño cycle and experienced much more severe winter wind and wave climates than "average". Wind and wave roses for the first four seasons, from winter 1998-99 through until summer 2000 are shown in Figures 5.5 and 5.4.

For these reasons it was decided that the wave input reduction should capture the distinct wave climates of the 10 individual seasons from "Winter 1998" (November 1998 to April 1999) through until "Summer 2003" (May to October 2003). The chronology of wave events occurring during each of these seasons was judged to be unimportant. Correlations between forcing processes *within each season* were also considered. Figure 5.3 shows an example time series of wind and wave conditions for the winter of 1998-1999 and the resulting correlations between wind, wave, river, and low-frequency water level fluctuations per season are indicated in Figure 5.6. Moderately strong correlations are observed between wave and wind north components, and between wave height and river discharge and low-frequency water level fluctuations. As both wind and waves are expected to be significant in terms of residual sediment transport, any reduction in the wind and wave input series must also capture this partial correlation. Similar levels of correlation exist between waves and river discharge and low-frequency water levels however, as these latter processes were not expected to play an important role in determining residual sediment transport in Willapa Bay, the correlations between these processes and wave forcing are ignored. These somewhat subjective judgements should only be regarded as a baseline against which sensitivity tests can be performed. The correctness of these judgements was tested by performing sensitivity simulations with the forcing processes altered and is discussed in Section 5.3.4.

Manual wave class selection

Within each season, two methods of wave input reduction were employed. The first method, used in the baseline accelerated simulation, required "engineering judgement" to manually divide the wave climate of each season into a number of wave classes based on wave parameters. The number and arrangement of wave classes selected varied from season to season, depending on the range of conditions recorded and a weighted mean wave condition and probability of occurrence was defined for each wave class for each season. To assist with this task a simple wave climate analysis tool was developed in Matlab. The tool plots the distribution of wave records (every 60 minutes) for each season in two-dimensional space based on the significant wave height (H_s) and peak wave direction (D_p). The proposed wave class boundaries are then superimposed on the wave data and the weighted centroid coordinates (H_s, T_p, D_p) of the recorded data in each wave class are also displayed. Within each wave class, data points are weighted proportional to $H_s^{2.5}$ as this exponent roughly corresponds with the CERC formula for longshore transport and is frequently used to estimate the "morphological impact" of waves. To assist with determining the location of the wave class boundaries, a number of ad-hoc statistics were also computed for each wave class. These include an estimate of the relative total morphological impact of waves in the wave class, estimated according to

$$M_c = p_c H_{s,\text{rep}}^{2.5} = \frac{\sum_{i=1}^{N_c} H_{s,i}^{2.5}}{N_s} \tag{5.17}$$

and the "morphological scatter" in the wave class, according to

$$\Delta_m = M_c \frac{\sum_{i=1}^{N_c} \sqrt{\left[|H_{s,i} - H_{s,\text{rep}}|^{2.5}\right]^2 + [\Theta_i - \Theta_{\text{rep}}]^2}}{N_c} \tag{5.18}$$

where N_c and N_s are the number of wave records in the wave class and season respectively, $H_{s,i}$ and Θ_i are the individual significant wave height and direction records, and $H_{s,\text{rep}}$ and Θ_{rep} are the representative significant wave height and direction for the wave class. These statistics for each wave class are also computed as a percentage of the total morphological impact and scatter for the entire season. Wave class boundaries were manually modified until the morphological impact and morphological scatter of each wave bin was similar. Example output for two seasons from the wave climate analysis tool are shown in Figure 5.7.

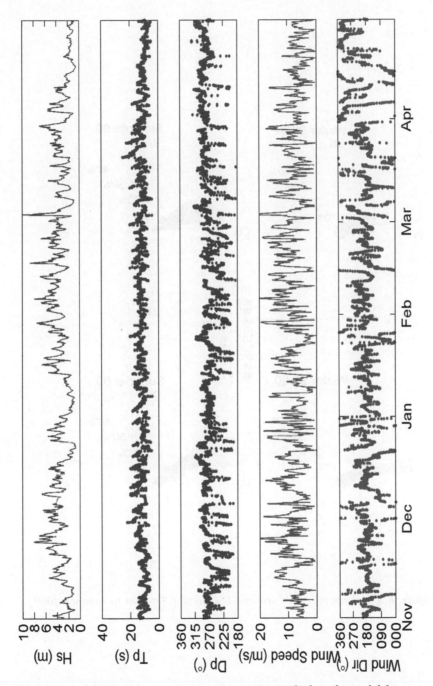

Figure 5.3 – *Wind and wave forcing time series applied to the model for season 1, winter 1998-99.*

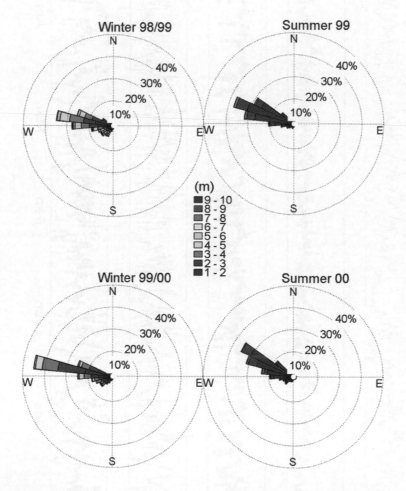

Figure 5.4 – *Wave roses for seasons 1–4, winter 1998-99 to summer 2000.*

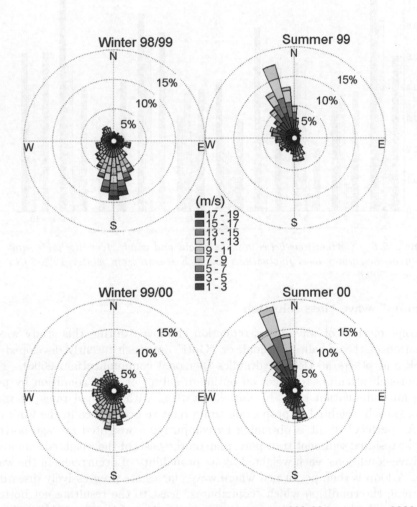

Figure 5.5 – *Wind roses for seasons 1–4, winter 1998-99 to summer 2000.*

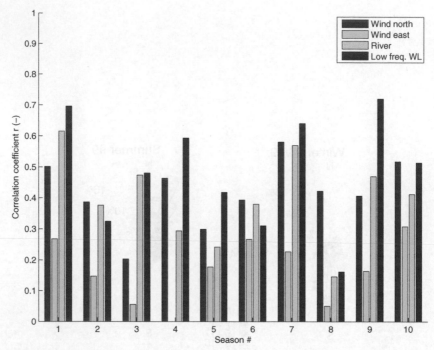

Figure 5.6 – *Correlations between wave height and wind, river discharge, and low-frequency water level fluctuations for each season from winter 1998-99 to summer 2003.*

"Optimum" wave class selection

The second method of wave input reduction employed during this study was a modification of the so-called optimal, or "Opti" approach recently developed by Roelvink and others at Delft Hydraulics (personal communication, 2006). The Opti approach attempts to select an optimum subset of wave condition by performing an "elimination race" by initially starting with a broad range of wave conditions, each weighted in proportion to its relative occurrence in the wave climate. A "perfect" result is obtaining by computing a weighted average bottom change, or residual sediment transport, map on the basis of the weighted sum of all of the wave conditions, each weighted by its probability of occurrence in the wave climate. A loop is then performed where wave classes are progressively discarded by dropping the condition which "contributes" least to the resulting net bottom change or residual sediment transport. The weighting of the dropped condition is then added to the most closely correlated remaining condition, and an inner loop is performed where random factors between zero and two are applied to the weightings of up to 3 of the most closely correlated remaining conditions and the RMS error between the new weighted sum of bottom changes, or sediment transports, and the "perfect" result is computed. This loop is iterated many times and the array of randomly assigned class weights which minimises the RMS error with the perfect solution is adopted. The condition which contributes least to

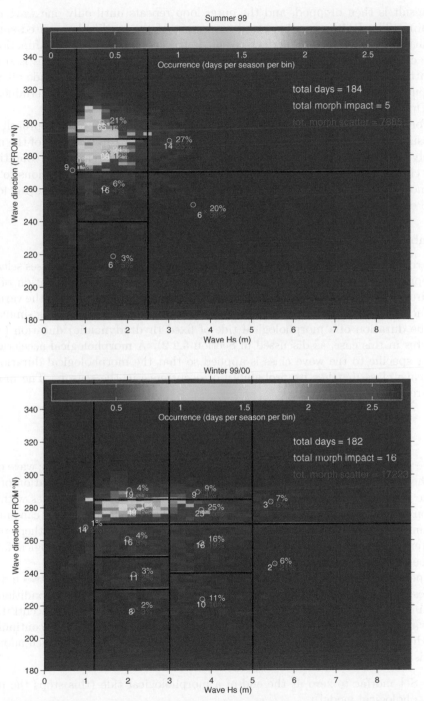

Figure 5.7 – *Example use of the wave schematisation tool to assist selection of wave classes for two seasons: summer 1999 (upper panel) and winter 1999-2000 (lower panel). Black lines show boundaries of proposed wave classes. Red cross indicates the weighted centroid of each wave class.*

the result is then dropped, and the outer loop repeats until only one wave class remains. The RMS error, bias, and covariance of the modified weighted sum of conditions relative to the "perfect" result is tracked and plotted as a function of wave classes remaining. Selection of the "optimum" set of wave classes is simply a matter of deciding what level of RMS error offers a reasonable trade-off with computational effort and selecting the set of classes and weightings determined by the Opti process for this level of accuracy.

For this study the only modification from the "standard" Opti approach was to assign the random weightings to just a limited number (up to three) of the most closely correlated remaining conditions, rather than to all remaining conditions. This change was made as it was found that limiting the random variations to just the most closely correlated wave conditions resulted in consistently achieving lower RMS error values for the same number of random trial iterations.

Combination with morphological acceleration

Regardless of which technique is used to select the wave classes, the classes selected to represent each season must be turned into a morphological simulation of the required actual season length. This is extremely simple to achieve with the variable morfac approach to morphological acceleration. Each wave condition is simulated for the duration of a morphological tide of fixed (hydrodynamic) duration (1490 minutes in this case, as discussed in Section 5.2.2). A morphological acceleration factor specific to the wave class is applied so that the morphological duration of the wave class matches its probability of occurrence in the season. The morfac required is computed by

$$f_{\mathrm{morfac}} = \frac{p_c \times \text{season duration}}{T_{\mathrm{morph\ tide}}}$$

where p_c is the probability of occurrence of conditions falling within this wave class for the season in question, and $T_{\mathrm{morph\ tide}}$ is the (hydrodynamic) duration of the of the representative morphological tide. The order of the wave classes within a season is randomly assigned. In this manner the hydrodynamic model can simply be run for the required number of morphological tides, one after another, and a different offshore wave boundary condition and corresponding morphological factor can be applied to each successive tide.

Special care needs to be taken when changing morfac values between wave classes, especially when simultaneously changing the wave boundary condition, as it is important that approximately the same suspended sediment concentrations exist at the start and end of a morfac value, otherwise significant discontinuities in sediment mass will occur, as discussed in Section 5.2.1. The approach adopted in this study when changing wave conditions was to

1. Set morfac to zero at the end of a morphological tide (this stops the morphological model),

2. Alter the required hydrodynamic model boundary conditions (wind, waves, etc) over a period of 60 minutes to avoid shocking the models,

3. Allow the hydrodynamic and wave models to stabilise to the new boundary conditions over the following 685 minutes

4. Set morfac to the appropriate value for the new wave class, (this restarts the morphological model),

5. Compute hydrodynamics, sediment transport, and accelerated morphological change through exactly one morphological tide,

6. Set morfac to zero to halt the morphological model.

7. Repeat the above steps until all wave classes in all seasons have been completed.

The above approach ensures that the start and end of individual morfac values occur at identical times in the harmonic tidal cycle and with identical wave conditions. This should ensure that suspended sediment concentrations are similar at the two times and any discontinuity of sediment mass is minimised. The method could be improved further by ensuring that the morfac transitions occur at around slack water, when suspended sediment concentrations are minimum.

5.2.4   Wind input reduction

The key challenge to be faced when schematising wind input to a morphodynamic model is that it is frequently (partially) correlated with offshore wave conditions. This is the case at Willapa Bay where wind and wave records gathered at the nearby buoy offshore of the Columbia River mouth indicate that there is a partial correlation between wind and waves (Figure 5.6). In this study a simple, but commonly adopted, approach was followed in order to capture this partial correlation. For each wave class defined in the schematised wave climate, the mean wind *stress* components were computed, based on the recorded wind speed and direction occurring during each wave record that fell within the boundaries of the defined wave class. These mean wind stresses varied from wave class to wave class and from season to season. An equivalent, representative, wind was then computed which would have the same wind *stress* components as the mean recorded wind stress for each wave class. This representative wind was then applied to the hydrodynamic model during each wave class. As discussed above, the transition from one wind/wave class to the next must be handled carefully to avoid shocking the hydrodynamic simulation and minimising the loss of continuity of sediment mass.

5.2.5   Other processes

Other forcing processes that need to be simplified will vary from location to location. In the case of Willapa Bay, two further processes need to be simplified in order to run accelerated morphological simulations. These are the discharge of the Willapa and Naselle rivers into Willapa Bay, and the slowly-varying, seasonal, fluctuation in coastal water levels. Both of these processes show significant correlation with offshore wave height, even within each season (Figure 5.6) however, as neither process is expected to be particularly important to the morphology of the estuary entrance, very simple input reductions have been made. In each case, the mean level of the process (river discharge or water level) was determined for each season and this is applied as a constant boundary condition for each season. In order to avoid shocking the hydrodynamic model, a linear ramp of 3 hours duration is used to transition the boundary conditions at the start of each new season. For the accelerated simulations this transition occurs while morfac is set to zero. These additional processes could easily have a more detailed input reduction performed, which would capture the correlation with other processes, by simply determining the mean level of each process *per wave class* in a similar manner as described for wind forcing.

5.2.6 Validation of acceleration techniques

In order to quantify the errors introduced into a five year morphological simulation of the entrance to Willapa Bay by the use of the morphological acceleration and input reduction techniques described here this study first performs a series of brute-force simulations in which forcing processes are individually schematised, followed by a brute-force simulation with all forcing processes schematised. An accelerated simulation is then performed which utilises all schematised forcing processes in addition to the morfac morphological acceleration method.

Brier skill scores

The results of the brute-force and accelerated simulations are compared using Brier Skill Scores (BSS) which, as discussed by Sutherland et al. (2004), provide an objective quantification of the skill of a model, where skill is defined to be the accuracy of a model prediction *relative to a baseline prediction*. In general, the formulation for the BSS is

$$BSS = 1 - \frac{\left\langle (Y - X)^2 \right\rangle}{\left\langle (B - X)^2 \right\rangle}$$

where Y is a prediction, X is an observation, and B is a baseline prediction against which the model skill will be evaluated. $\langle \rangle$ denotes taking an arithmetic mean.

In the present application, for judging the decrease in skill caused by introducing the various input reduction and morphological acceleration techniques, the BSS is set up as follows

$$BSS = 1 - \frac{\left\langle (\Delta h_{\text{predicted}} - \Delta h_{\text{benchmark}})^2 \right\rangle}{\left\langle (\Delta h_{\text{benchmark}})^2 \right\rangle}$$

where $\Delta h_{\text{predicted}}$ is a spatial map of model predictions of sedimentation and erosion, $\Delta h_{\text{benchmark}}$ is a spatial map of "perfect" benchmark brute-force model predictions of sedimentation and erosion. This effectively sets the baseline result B

for comparison of the model accuracy equal to the initial model bathymetry, or a baseline model of zero morphological change.

When set up in this manner, a BSS of 1.0 implies that the five year morphological change predicted by a model simulation is perfect – i.e. it is identical to the "perfect" benchmark five year model simulation using full time series of all forcing processes. A BSS of 0.0 would imply that the morphological change predicted by the model simulation is no better than the baseline model of assuming no morphological change. The BSS can become negative if the model result is less accurate than the baseline model. van Rijn et al. (2003) proposed a qualification of BSS results for morphodynamic studies, as repeated in Table 5.2.

Although the BSS provides a single number to quantify the skill of a model result, it does not help provide qualitative information on how, why, when, or where a model result deviates from the target. To assist with this, separate BSS have been computed for four distinct regions of the model, as well as for the model as a whole. This allows one to quickly assess where changes in model skill are occurring and therefore help deduce why the change occurs. The four regions selected: "Outer Coast", "Entrance Shoal", "Entrance Channel", and "Inner Estuary" are shown in Figure 5.8. The division between the Entrance Shoal and Outer Coast regions is located at the -10m MSL depth contour. Time histories of cumulative sediment transport through selected predefined control sections are also used to help identify at what point in time model results diverge. The location of the control sections defined in the Willapa Bay model are shown in Figure 5.1 on page 151

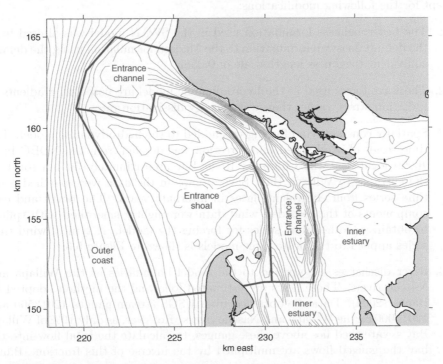

Figure 5.8 – *Brier Skill Score regions.*

Table 5.2 – Qualification of BSS error ranges for morphodynamic studies (van Rijn et al., 2003).

Qualification	BSS
Excellent	$1.0 - 0.8$
Good	$0.8 - 0.6$
Reasonable/fair	$0.6 - 0.3$
Poor	$0.3 - 0$
Bad	< 0

Brute-force morphological simulations

The hydrodynamic, wave, and sediment transport models used for the investigation of input reduction and morphological acceleration techniques are nearly identical to the WBAY model described in detail in Chapter 4 of this thesis. Slight changes were made to the model as the present study is performed with a later version of the Delft3D system than that used in Chapter 4. In the later version the module interfacing between Delft3D-FLOW and SWAN had been upgraded to provide a simpler "online" coupling between waves and hydrodynamics. This upgrade resulted in changes to the interpolated bathymetry passed from the hydrodynamic model to SWAN and required a slight recalibration of the SWAN wave model.

The brute force simulations use the WBAY model as described in Section 4.3.4 except for the following modifications:

1. The bed roughness formulation used in the SWAN model was altered from the default Jonswap formulation to the Madsen formulation using the default equivalent roughness length scale of 0.05m.

2. The wave forces used in the hydrodynamic model are based on gradients in radiation stress, rather than wave energy dissipation rates.

3. Spatially constant wind forcing was added to the hydrodynamic model. The time series of wind speed and direction was derived from the NDBC buoy 46029 located at the Columbia River Bar approximately 50 km south of the entrance to Willapa Bay. Several gaps in the data were patched using time series from equivalent dates in alternate years. The north and east components of the recorded wind data were lightly smoothed with splines to obtain smoother time series for forcing the model. A typical wind time series applied to the brute force models is shown in Figure 5.3.

4. River discharges are applied to the model boundaries at the Willapa and Naselle rivers. These flows are estimated following the method adopted by Banas (2004). Data are obtained from USGS stream gauges 12013500 and 12010000 on the Willapa and Naselle rivers. However, as only 19% of Willapa Bay's watershed lies above these gauges, to calculate the total flow into the bay the gauged flows are multiplied by the inverse of this fraction. Banas

reports that the resulting estimate is consistent with the annual mean value reported by NOAA/EPA (1991) to within 20%.

5. The harmonic offshore water level boundary condition in the WBAY model was replaced with a time series of water levels derived from a blend of tidal harmonics and low frequency measured water levels at Toke Point in Willapa Bay. This was achieved as follows:

   (a) Model simulations were used to estimate independent correlations between offshore wave height and water level setup at Toke Point and between the east component of the wind velocity and water level setup at Toke Point. These correlations were used to produce estimated time series of computed wave and wind setup at Toke Point for the period 1998 - 2003. The time series were then low-pass filtered with a cutoff frequency of 1/36 hours.

   (b) The measured water levels at Toke Point for the duration of 1998 - 2003 were low pass filtered with a cutoff frequency of 1/36 hours to remove the tidal signal.

   (c) The time series of estimated low frequency water level fluctuations attributable to wind and wave setup (which should be produced in the hydrodynamic model) were subtracted from the measured low frequency water level fluctuations at Toke Point. This leaves a time series of low frequency water level fluctuations unable to be resolved by the model.

   (d) Time series expansions for the period 1998 - 2003 were produced from the astronomical constituents applied at the western boundary of the WBAY model.

   (e) The time series of low frequency water level fluctuations were added to the time series of the astronomical constituents.

   (f) The resulting composite time series were applied at the western boundary of the model. Once the wind and wave forcing are also applied to the model domain the model reproduces the total water level variation (low and tidal frequency) observed at Toke Point with an RMS error of 0.12m over the 5 years simulated.

The brute-force model is run for the period from midnight on 31 October 1998 to midnight on 1 November 2003. The first 24 hours of simulation, through until midnight on 1 November 1998, is a hydrodynamic spin-up period and no morphological change takes place. The time step used in the model is 1 minute and the wave field is updated every 15 minutes. Each brute force simulation takes approximately 750 hours (31 days) to run on a single 3.2GHz processor. The computational effort is approximately evenly split between the Delft3D-FLOW and SWAN models. A series of seven brute force simulations are performed. The characteristics of the individual simulations are given in Table 5.3. By contrasting the results of these simulations the impact of each of the input reduction techniques on the computed morphological change can be isolated. The final simulation (bf7) containing all simplified forcing processes provides insight into the non-linear interaction between the various input reductions, and also provides the baseline for

Table 5.3 – *Characteristics of brute force simulations*

Simulation Code	Characteristics
bf1	Benchmark simulation. Uses full time series for all processes.
bf2	Simplified low frequency water level at offshore boundary.
bf3	Simplified river discharge time series.
bf4	Simplified wind forcing time series.
bf5	Simplified wave forcing time series.
bf6	Simplified tidal boundary condition time series.
bf7	All forcing time series simplified.

comparison with the first accelerated simulation which uses identical reduced input time series, but also uses the variable morfac approach to increase the morphological time step and reduce the duration of the simulation to a more manageable level.

The simplified input conditions used in the brute-force simulations are produced as described in Sections 5.2.2 to 5.2.5 and, when viewed on a morphological time scale, are nearly identical to the simplified forcing processes used in the accelerated simulations described below. Slight differences are unavoidable as the manner in which the transitions between conditions occur during times when morfac = 0 in the accelerated simulations is impossible to reproduce in the brute force simulations without shocking the hydrodynamic model. Short, three hour, transitions are applied in the brute force simulations with simplified forcing series. At a morfac of 1.0 these short transitions will not cause any significant error compared to the "instantaneous" transitions achieved in the accelerated simulations.

Accelerated morphological simulations

The hydrodynamic, wave, and sediment transport models used for the accelerated simulations are identical to the models used for brute force simulation *bf7*, with the exception of the altered time-scale for the forcing and boundary conditions and the use of morfac, the morphological acceleration factor. The absolute dates and times used in the hydrodynamic model are not relevant as all of the forcing processes are synthetic, simplified, input time series and are defined relative to the model start date. The first 12 hours of simulation is used for spinning up the hydrodynamics, waves, and sediment transport. Accelerated morphological developments then start at midnight on 1 November 1998, in an identical manner to the brute force simulations.

The morphological tide used in the accelerated simulations is a compound M_2 + C_1 double tide, as described in Sections 5.2.2 and 5.3.2. The frequencies of the M_2 and C_1 components are altered very slightly to make the morphological tide periodic at exactly 1490 minutes (24 hours 50 minutes) in order to keep output writing and input transition times exact multiples of the simulation time step. As discussed in Section 5.3.3, each winter season consists of 12 wave classes and each summer season 7. Thus 19 wave classes are required for each year simulated. The order of wave classes within a season is random, however for the first comparison simulation the same random sequence as used in the brute force simulation *bf7* is retained. Sensitivity to rearranging the order of the wave classes within the seasons will be investigated separately. Using just one morphological tide per wave class requires a maximum morfac of 77.45 for the most commonly occurring wave class with a 1.5m significant wave height. Morfacs in the range of 10 to 20 are required for wave classes with 3 - 4 m high waves when morphological developments are expected to be more rapid. Based on previous experience this range of morfac values seems reasonable. If individual values were judged to be too high they could be lowered by simply allowing two or more morphological tides for each of the affected wave classes.

When wave conditions change between classes, morfac is temporarily set to zero to avoid polluting the morphological model with sediment transport occurring during transients in the hydrodynamic model. In this study a relatively long relaxation period of 12 hours is allowed between wave classes. The wave and sediment concentration fields adapt much more quickly than this, however the along-coast current driven by the north component of the wind forcing takes considerable time to adapt to the new wind stress associated with each new wave class. The long-coast current doesn't fully stabilise even within 12 hours, but this was judged to be a reasonable compromise. In reality winds rarely blow consistently enough to enable the long-coast current to reach equilibrium with the wind stress, so this is one area of the input reduction which can be expected to be imperfect, however long the hydrodynamic simulation was allowed to spin up at the start of each wave class. Another way of looking at this is that because of the inertia in the offshore wind-driven current and the non-linear relationship between wind stress and current velocity, the chronology of wind forces *is* important in determining the strength of the long-shore current that develops and this phenomenon is not captured in the present wind input reduction method. If this aspect of the modelling is found to perform particularly poorly then an improved method of wind input reduction which respects the (typical) chronology of wind velocities in each wave class may need to be investigated.

As each wave class effectively requires 1.5 morphological tides of hydrodynamic simulation, the total duration of the accelerated simulation is 1.5×19 tides per year $\times$ 5 years = 142.5 tides in length (plus the initial spin-up) amounting to a total hydrodynamic simulation duration of approximately 148 days, or 8% of the duration of the brute force simulation.

A series of accelerated simulations is performed in order to investigate the sensitivity of the model result to the details of the input reduction approaches used (Table 5.4). The first simulation (*ac1*) is intended as a test of the error introduced by using the variable morfac morphological acceleration technique. This simulation uses identical simplified inputs as brute force simulation *bf7* and should, if the variable morfac acceleration technique worked perfectly, produce identical results. Subsequent simulations test alternative approaches to schematising the wave and wind climates (*wr1 – wr6*) or morphological tide (*tr1 – tr5*). The results of these simulations will be compared directly back to the benchmark brute force simulation *bf1* to compare the skill of these simulations with the wave and tide input reduction

Table 5.4 – Characteristics of accelerated simulations

Simulation ID	Characteristics
Baseline accelerated simulation	
ac1	Identical input reduction to brute force simulation *bf7*.
Wind input reduction series	
wr1	Different random order of wave classes within each season.
wr2	No chronology of years.
wr3	No chronology or seasonality.
wr4	Reduced number of wave classes.
wr5	Wave classes selected using "Opti" with mean wind.
wr6	Wave classes selected using "Opti" with seasonal wind.
Tide input reduction series	
tr1	Morph. tide = $1.09 \times M_2 + C_1$.
tr2	Morph. tide = $1.00 \times (M_2 + C_1)$.
tr3	Morph. tide = $1.20 \times M_2$.
tr4	Morph. tide = $1.12 \times M_2 + C_1$.
tr5	Morph. tide = $1.08 \times M_2$.

methods used in simulation *ac1*.

5.3 Results

5.3.1 Benchmark simulation

The benchmark brute force simulation *bf1*, driven by complete time series of forcing processes, shows behaviour which is qualitatively typical of the observed behaviour of the entrance to Willapa Bay (Figure 5.9). The outer end of the main channel shows a tendency to first rotate to the north, then switch abruptly to the south, before re-stabilising with a more central orientation. This type of behaviour is characteristic of mixed-energy, tide-dominated inlets and can be explained by the "outer channel shifting" model of sediment bypassing described by FitzGerald et al. (2000). Temporal animations of these changes indicate that they appear to be seasonal, with relatively little change occurring during summer months. Significant erosion occurs in the main channel, with the channel depth increasing during the early years of the simulation again, broadly in line with observations. The large quantity of sand eroded from the main channel is predominantly exported from the bay (Figure 5.10) with the bulk depositing on the outer north-west edge of the ebb tidal delta.

Northwards-directed longshore sediment transport along the outer coast is considerable, occurring almost entirely during winter months (Figure 5.11. Residual sediment transport through the Leadbetter control section, located at the southern end of the estuary mouth, is rather small however as the northwards-directed sediment transport at depths below about -10m MSL is balanced by southwards-directed transport in shallower water. This behaviour has also been previously documented by Ruggiero et al. (2005).

Residual sediment transport in the estuary entrance is dominated by a circulation of sediment inwards over the shoals and shallower "Mid" and "South" channels (Figure 5.12) and outwards through the deep main channel. Net export of sediment in the main channel is computed to be almost 20 million cubic metres in 5 years, although approximately half of this is offset by import over the entrance shoals (Figure 5.13). A strong export of sediment also occurs in the Nahcotta channel, caused by the residual circulation of water and sediment returning from the shoals to the main channel.

Inside the estuary, away from the strong residual circulations in the entrance, residual sediment transports are one to two orders of magnitude smaller. Export-directed residual sediment transport through section "Mid1" in the centre of the southern arm of Willapa Bay is significant at 1.6 million m^3 in 5 years. Residual export-directed transports through the "South1" and "South2" control sections are a further order of magnitude smaller. The "Toke" control section displays a small flood-directed residual transport, and is the only section to display this behaviour away from the wave-dominated entrance shoals.

Despite the good qualitative agreement between the morphological model and the observed behaviour of the entrance to Willapa Bay, quantitative comparison of modelled and measured sedimentation and erosion patterns indicates that the model has little quantitative skill. The BSS for the various regions of the benchmark brute-force simulation are shown in Table 5.5 and indicate that the only region of the model *not* regarded as having "bad" predictive skill is the Outer Coast region. This is something of a surprise, especially for the Entrance Channel region, which visually appears to capture the main measured morphological change patterns (Figure 5.15). This is of concern for predicting the morphological change of Willapa Bay, but not for the present study as what concerns us is the ability to consistently reproduce the benchmark model results with simulations incorporating input reduction and morphological acceleration. If the model result displayed none of the characteristics of the natural estuary, or showed little sensitivity to variation in forcing processes, then the usefulness of the following results would be questionable. However the process calibration carried out in Chapter 4 of this thesis and the sensitivity analysis performed in Sections 5.3.8 and 5.3.9 of the present chapter show that the hydrodynamic model is well validated against measured hydrodynamics and the morphodynamic model is sensitive to subtle change in forcing conditions. Therefore the simulation *bf1* result described here can be used as a benchmark simulation where all forcing processes are known and fully described, notwithstanding the errors that may exist in the initial conditions and/or physical process formulations.

5.3.2   Tidal input reduction

Subtracting the result of simulation *bf1* from simulation *bf6* isolates the error introduced by simplifying the full tidal signal to the morphological double tide. It is important to bear in mind that introducing the morphological tide not only

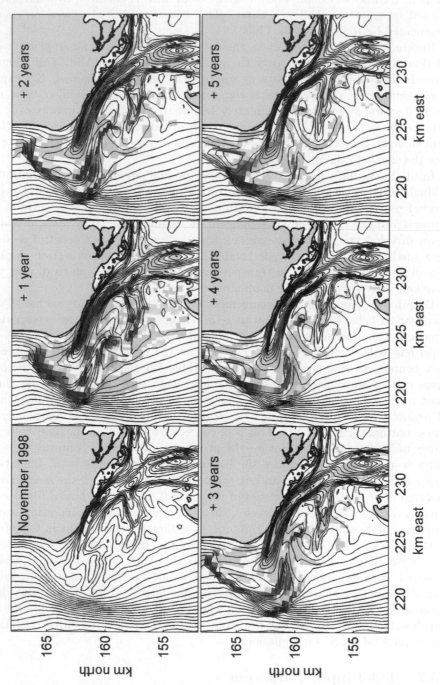

Figure 5.9 – *5 years of morphological change predicted by benchmark brute force simulation bf1. Red and blue colours indicate areas of annual sedimentation and erosion respectively.*

Table 5.5 – *BSS for regions of the benchmark simulation bf1 against observations, relative to a baseline prediction of zero morphological change.*

Model region	BSS
Outer Coast	0.23
Entrance Shoal	−0.11
Entrance Channel	−0.80
Inner Estuary	−0.40
Overall	−0.47

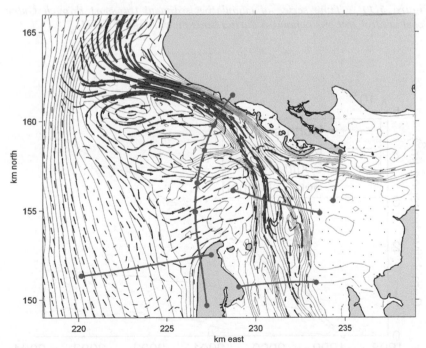

Figure 5.10 – *Curved vector representation of residual sediment transport patterns computed during benchmark simulation bf1. Location of control sections shown in magenta.*

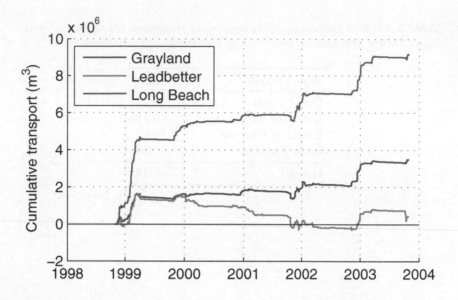

Figure 5.11 – *Time series of cumulative sediment transport through Outer Coast control sections computed during benchmark simulation bf1.*

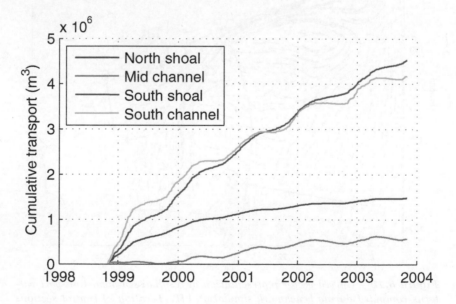

Figure 5.12 – *Time series of cumulative sediment transport through Entrance Shoal control sections computed during benchmark simulation bf1.*

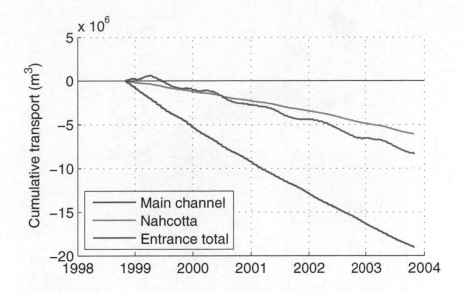

Figure 5.13 *– Time series of cumulative sediment transport through Entrance Channel control sections computed during benchmark simulation bf1.*

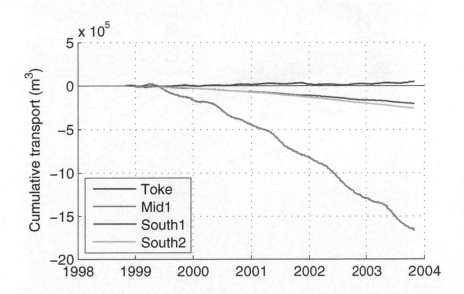

Figure 5.14 *– Time series of cumulative sediment transport through Inner Estuary control sections computed during benchmark simulation bf1.*

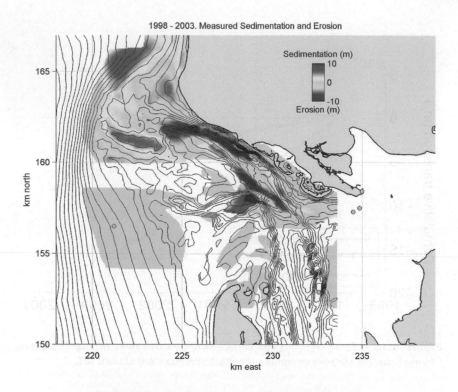

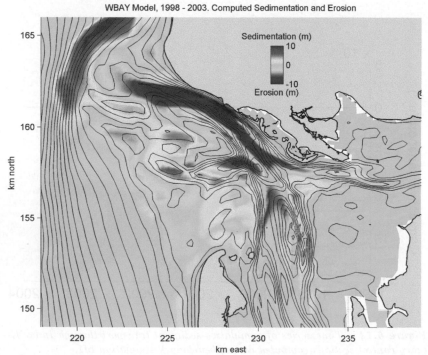

Figure 5.15 – *Patterns of sedimentation and erosion 1998 to 2003. Measured (upper panel) and computed by benchmark simulation bf1 (lower panel).*

changes the residual sediment transport due to tide, but also due to the interaction of tide and all other forcing processes. As residual sediment transport in the Willapa Bay model is significantly influenced by at least waves and wind in addition to tide, the disruption of the relative phasing of these forcing processes will also have an effect. Error introduced into simulation *bf6* is greater than would be the case if tide was the only significant forcing process.

Figure 5.16 shows a map of the magnitude of the error in the residual sediment transport vector as a fraction of the magnitude of the benchmark residual sediment transport. Areas with very low residual sediment transport rates and areas where the error is less than 20% have been masked out of the figure. The remaining coloured areas highlight locations where the introduction of the morphological tide has caused a significant error in a significant residual sediment transport vector. The vast majority of the model area is not significantly affected by the use of the morphological tide. However, ignoring isolated single grid cells, two classes of location do appear to be affected:

1. Offshore, beyond the outer end of the channels through the ebb shoal. In these regions the benchmark residual sediment transport field shows strong gradients, divergences, and circulations. They are areas of rapid and spatially variable morphological change. It is possible that errors in these locations may be caused by the loss of relative phase between tide and wave forcing processes. The error introduced into the residual sediment transport vectors will manifest itself as local spatial offsets in computed morphological change patterns.

2. In channels well inside the estuary with little residual flow. Two clear examples in Figure 5.16 are near the Toke and Mid1 control section locations. The temporal development of the cumulative sediment transport through the Inner Estuary control sections is shown in Figure 5.17. It is clear that use of the morphological tide causes a consistent and progressive error in the residual transports through both the Toke and Mid1 control sections. In both cases use of the morphological tide has resulted in additional ebb-directed residual transport. A closer view of the error in the residual sediment transport in the vicinity of the Toke control section can be seen in Figure 5.18. It is clear that this is an area with a complex divergence of residual transports, with a relatively strong ebb-directed residual transport of the south side of the channel and a weaker flood-directed residual on the north side. The morphological tide appears to capture the stronger ebb-directed residual well, but under estimates the flood-directed residual.

The impact of introducing the morphological tide on predicted morphological change patterns is evaluated by computing BSS for the four model regions and the model as a whole (see column *bf6* in Table 5.6). The skill of the model is not significantly degraded from perfect in any model region (the minimum BSS is 0.97 for the entrance shoals).

In summary, the introduction of the morphological tide has been shown to introduce insignificant error into the 5 year morphological simulation, with just very minor impact on morphological change predicted on the outer coast and entrance shoals in the vicinity of strong gradients in residual sediment transport at the outer ends of the main tidal channels. While this result is generally very acceptable, the systematic errors observed in the relatively minor residual transports through the Toke and Mid1 control sections are unwelcome. It is possible that the extra export residual transport through these sections could be caused by the amplification factor applied to the $M_2 + C_1$ tidal constituents, as discussed in Section 5.2.2. This hypothesis is tested using accelerated simulations, and is discussed in Section 5.3.9 below.

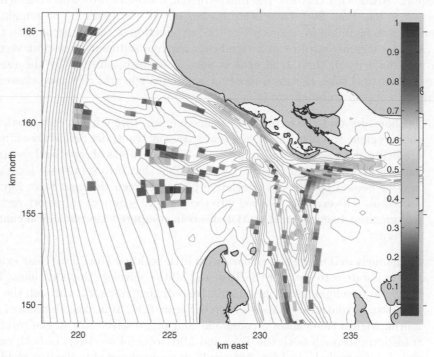

Figure 5.16 *– Map of error in residual sediment transport caused by use of the morphological tide. Colours show the magnitude of the error vector as a fraction of the magnitude of the true residual sediment transport vector.*

5.3.3 Wave input reduction

Simulation *bf5* contains the simplified wave climate. Again, it is important to note that at Willapa Bay both wind and waves have a significant impact on residual sediment transport and are partially correlated. Introducing just one simplified process into the brute force model destroys this correlation and is quite likely to have a greater detrimental effect than just that caused by the simplification of the waves in isolation. The map of errors caused by wave schematisation (Figure 5.19) shows that the majority of the error in residual sediment transport is located, as one would expect, offshore and to a minor extent over the entrance shoals. Inspection of the cumulative transport through the offshore control sections (Figure 5.20) reveals that the schematised wave climate is consistently under-predicting

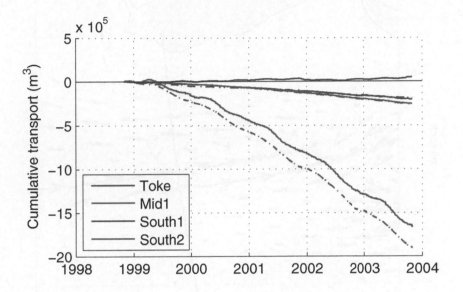

Figure 5.17 – *Time series of cumulative sediment transport through Inner Estuary control sections computed during benchmark simulation bf1 (solid lines) and simulation using morphological tide $1.08 \times (M_2 + C_1)$ bf6 (dash-dot lines).*

Table 5.6 – *Brier Skill Scores for the baseline accelerated simulation.*

Model	Brier Skill Score							
Region	bf2	bf3	bf4	bf5	bf6	bf7	ac1-bf7	ac1-bf1
Outer Coast	0.98	1.00	0.90	0.67	0.99	0.80	0.97	0.86
Entrance Shoal	0.99	1.00	0.96	0.96	0.97	0.98	0.99	0.97
Entrance Channel	1.00	1.00	1.00	0.96	0.99	0.99	1.00	0.99
Inner Estuary	1.00	1.00	1.00	1.00	0.98	0.98	1.00	0.98
Overall	0.99	1.00	0.97	0.92	0.98	0.96	0.99	0.96

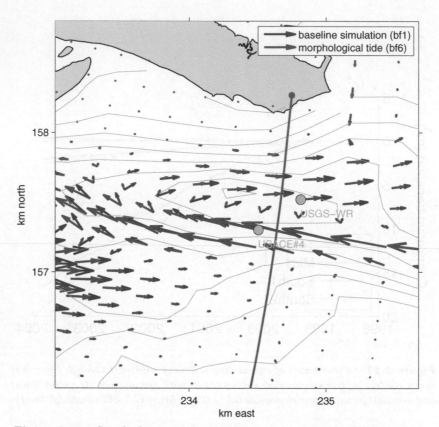

Figure 5.18 – *Residual transport vectors near Toke Point for benchmark simulation bf1 (blue) and simulation with morphological tide bf6 (red).*

the northwards net transport of sediment along the coast. The error is fairly consistent through all three control sections, with a consistent deficit in transport of approximately 1.5 million m³. Closer inspection of the time series shows that the deficit occurs primarily in the first winter (1998) which produces an abnormal amount of northwards transport in the benchmark simulation. In latter years the benchmark and schematised simulations track each other reasonably consistently.

Cumulative transports over the entrance shoals are much closer to the benchmark simulation (Figure 5.21) with errors within a couple of percent for transports through North shoal, South shoal, and South channel, and reaching 20 percent of the relatively small residual transport through Mid channel (Table 5.7). Wave schematisation causes the greatest errors recorded in the BSS in Table 5.6, with the BSS for the Outer Coast region dropping to 0.67. BSS for regions further into Willapa Bay are little affected by wave schematisation, with a lowest BSS of 0.96. While even the Outer Coast result is still rated as "Good" according to van Rijn, it should be possible to more accurately schematise the wave climate. This will be tested using accelerated simulations in Section 5.3.8.

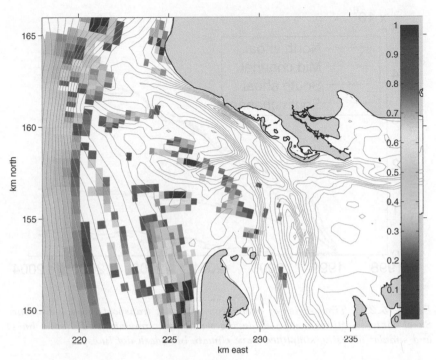

Figure 5.19 – *Map of error in residual sediment transport caused by use of the simplified wave climate. Colours show the magnitude of the error vector as a fraction of the magnitude of the true residual sediment transport vector.*

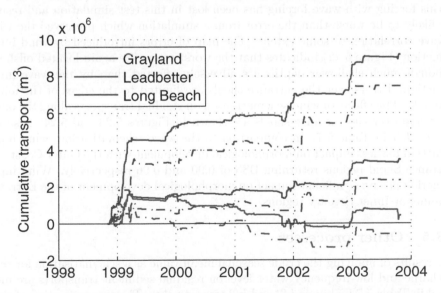

Figure 5.20 – *Time series of cumulative sediment transport through Outer Coast control sections computed during benchmark simulation bf1 (solid lines) and simulation using simplified wave climate bf5 (dash-dot lines)*

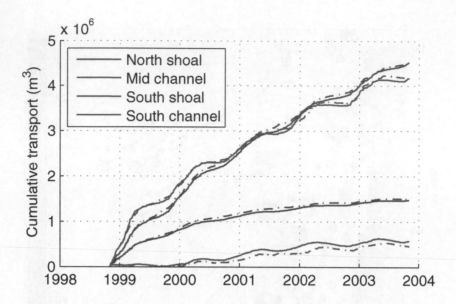

Figure 5.21 – *Time series of cumulative sediment transport through Entrance Shoal control sections computed during benchmark simulation bf1 (solid lines) and simulation using simplified wave climate bf5 (dash-dot lines).*

5.3.4   Wind input reduction

Simulation *bf4* contains the simplified wind forcing climate. Again, the correlation of this forcing with wave forcing has been lost in this test simulation and results are likely to be worse than the error from a simulation which preserved the wind – wave correlation to some extent. The map of errors introduced by wind input reduction (Figure 5.22) indicates that the worst errors are again located offshore, although much shallower depths are adversely affected than by the wave input reduction. Errors over the entrance shoals are limited to the edges of the minor channels. The effects of wind schematisation on residual transports along the outer coast and over the entrance shoals are shown in Figures 5.23 and 5.24. Also see column *bf4* in Table 5.7. In comparison to the wave schematisation, wind input reduction has less impact on morphological development with the Outer Coast and Entrance Shoal regions returning BSS of 0.90 and 0.96 respectively. Wind input reduction has no significant impact on morphological development of the Entrance Channel or Inner Estuary regions.

5.3.5   Other processes

The results of applying the crude seasonal mean value as a schematisation for river discharge and low-frequency water level on residual sediment transports are indicated in Table 5.7 (columns *bf3* and *bf2* respectively). The schematisation of river flows makes a very slight (< 10%) difference to the residual transports through the very sensitive Toke and Leadbetter control sections. The schematisation of

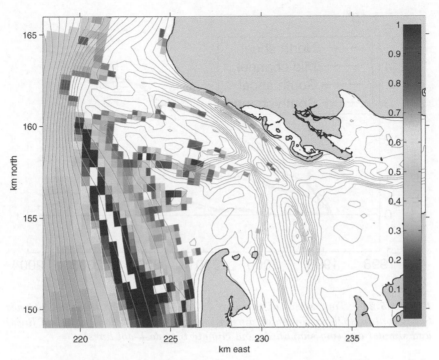

Figure 5.22 – *Map of error in residual sediment transport caused by use of the simplified wind climate. Colours show the magnitude of the error vector as a fraction of the magnitude of the true residual sediment transport vector.*

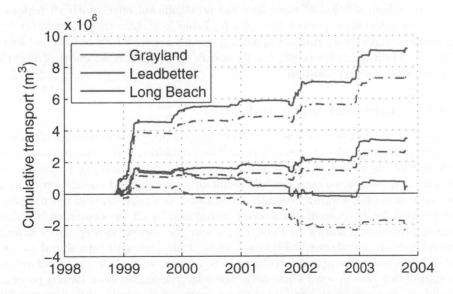

Figure 5.23 – *Time series of cumulative sediment transport through Outer Coast control sections computed during benchmark simulation bf1 (solid lines) and simulation using simplified wind climate bf4 (dash-dot lines).*

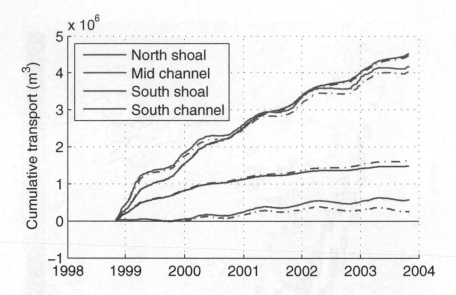

Figure 5.24 *– Time series of cumulative sediment transport through Entrance Shoal control sections computed during benchmark simulation bf1 (solid lines) and simulation using simplified wind climate bf4 (dash-dot lines).*

low-frequency water level fluctuations introduces a larger error, with significant errors introduced into transports through the Leadbetter, Mid channel, and Toke control sections the reasons for which are unclear. In terms of morphological impact, the schematisation of river flow has no significant effect with all regions of the model achieving a perfect BSS of 1.0 (Table 5.6). Low frequency water level fluctuations have a very minor impact but, with an overall BSS of 0.99 and lowest BSS of 0.98 on the outer coast, hardly justify spending further effort attempting to improve the schematisation.

5.3.6   Combined input reduction

Simulation *bf7* includes all the simplified forcing processes described previously, but still runs as a brute-force simulation. This simulation serves two purposes: first as a test of the extent of non-linear interactions between forcing processes, and second, as a baseline to compare the result of an accelerated simulation using all the same simplified forcing processes in addition to a time-varying morphological acceleration factor. A map of the error introduced by all the combined simplified forcing processes is shown in Figure 5.25 and the corresponding errors through control sections are shown in Figures 5.26 to 5.29. The error introduced by combined input reduction appears to be close to a linear sum of the errors introduced by simplifying each process separately. Some interaction between forcing processes is evident, however, with the Outer Coast region of simulation *bf7* having a BSS of 0.80 despite including both wind (BSS=0.90) and wave (BSS=0.67) simplifications. Inspection of Figure 5.25 confirms that the error in residual transport in offshore areas is less for the combined input reduction than for the individual simplified wind and wave simulations. Clearly, using an input reduction which retains some degree of the correlation between wind and waves is not as harmful as simplifying just wind or waves in isolation.

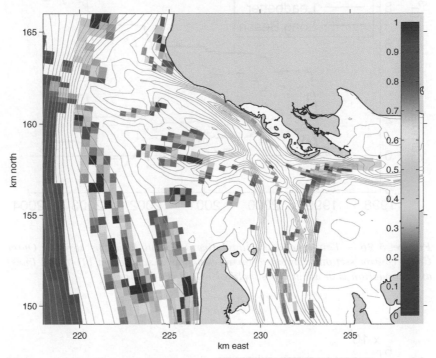

Figure 5.25 – *Map of relative error in residual sediment transport caused by the combined simplification of all forcing processes. Colours show the magnitude of the error vector as a fraction of the magnitude of the true residual sediment transport vector.*

5.3.7 Morphological acceleration

Comparing the result of the first accelerated simulation *ac1* with simulation *bf7* isolates the impact of introducing the variable morphological acceleration factor (morfac). The map of residual transport error (Figure 5.30) shows a scattering of minor impacts on the outer edge of the entrance shoals, in a pattern not dissimilar to that produced by the inclusion of the morphological tide. It is likely that these patterns are simply highlighting areas with particularly sensitive residual sediment transport patterns. The impact of the accelerated simulation on cumulative sediment transports through sections is shown in Figures 5.31 to 5.34. The deviation in final cumulative sediment transport is very slight ($< 6\%$ error for all sections except Leadbetter, Mid channel, and Toke) although large deviations during the elongated tidal cycles used by the simulation are clearly visible. It is important to recall that these intra-tide results are not meaningful in an accelerated simulation and should be ignored. To avoid this confusion, the cumulative transport plots

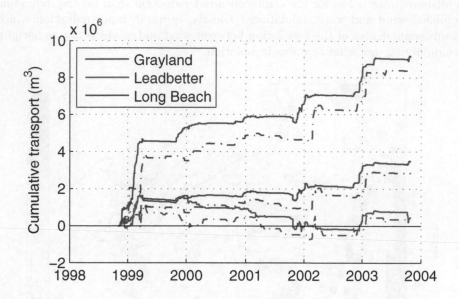

Figure 5.26 – *Time series of cumulative sediment transport through Outer Coast control sections computed during benchmark simulation bf1 (solid lines) and simulation with all inputs simplified bf7 (dash-dot lines).*

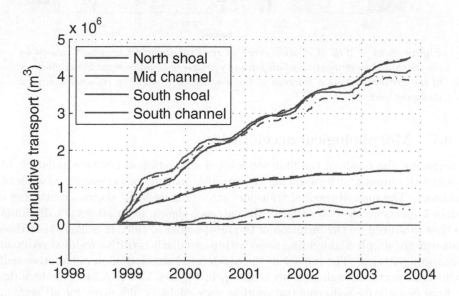

Figure 5.27 – *Time series of cumulative sediment transport through Entrance Shoal control sections computed during benchmark simulation bf1 (solid lines) and simulation with all inputs simplified bf7 (dash-dot lines).*

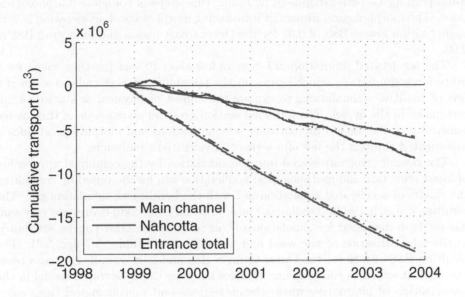

Figure 5.28 *– Time series of cumulative sediment transport through Entrance Channel control sections computed during benchmark simulation bf1 (solid lines) and simulation with all inputs simplified bf7 (dash-dot lines).*

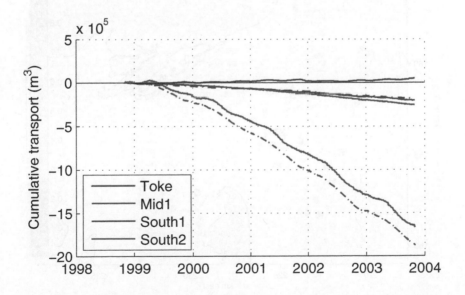

Figure 5.29 *– Time series of cumulative sediment transport through Inner Estuary control sections computed during benchmark simulation bf1 (solid lines) and simulation with all inputs simplified bf7 (dash-dot lines).*

could perhaps be better displayed by taking time steps of complete morphological tides. The morphological impact of introducing morphological acceleration is very slight, with a lowest BSS of 0.97 for the Outer Coast region and an overall BSS of 0.99.

The accelerated simulation *ac1* runs in less than 10% of the time taken for a brute-force simulation, which opens up the possibility of performing meaningful sets of sensitivity simulations to explore the impact of physical or statistical uncertainties in the model. The following section provides an example of this, as the sensitivity to the use of different wave schematisations and morphological tides is investigated through the use of a series of accelerated simulations.

The overall error introduced into the simulation by the combined approaches of input reduction and morphological acceleration can be computed by comparing the results of accelerated simulation *ac1* with the benchmark simulation *bf1*. The resulting errors in residual sediment transports and Brier Skill Scores are very similar to those discussed for simulation *bf7*, as most of the error can be attributed to the schematisation of the wind and wave climates (Tables 5.7 and 5.6). The resulting BSS of 0.86 for the Outer Coast region and 0.96 overall serve as a baseline against which to compare any changes made to the accelerated model in the investigation of alternative wave schematisations and morphological tides using the sequence of accelerated simulations described below.

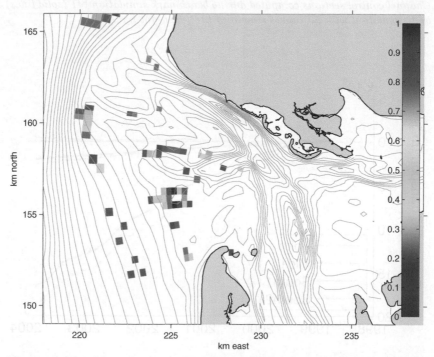

Figure 5.30 – Map of relative error in residual sediment transport caused by the introduction of morphological acceleration using variable morfac. Colours show the magnitude of the error vector (ac1-bf7) as a fraction of the magnitude of the residual sediment transport vector in simulation bf7.

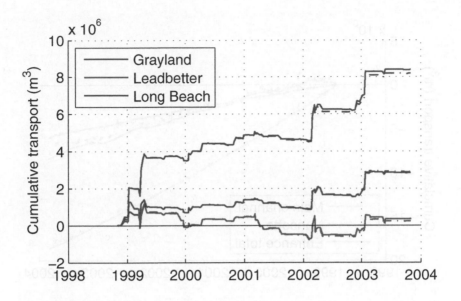

Figure 5.31 – *Time series of cumulative sediment transport through Outer Coast control sections computed during brute-force simulation with all inputs simplified bf7 (solid lines) and an accelerated simulation with identical inputs ac1 (dash-dot lines).*

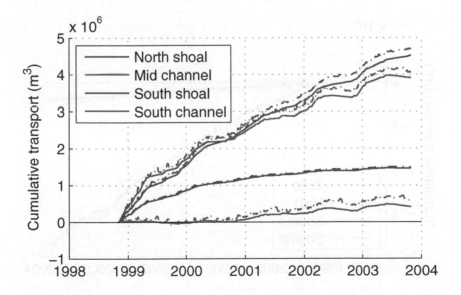

Figure 5.32 – *Time series of cumulative sediment transport through Entrance Shoal control sections computed during brute-force simulation with all inputs simplified bf7 (solid lines) and an accelerated simulation with identical inputs ac1 (dash-dot lines).*

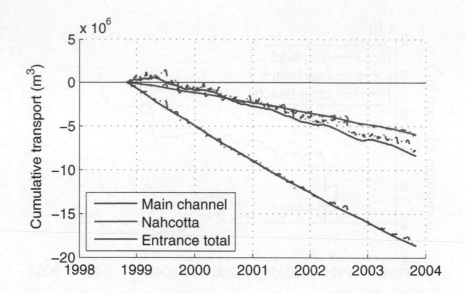

Figure 5.33 – *Time series of cumulative sediment transport through Entrance Channel control sections computed during brute-force simulation with all inputs simplified bf7 (solid lines) and an accelerated simulation with identical inputs ac1 (dash-dot lines).*

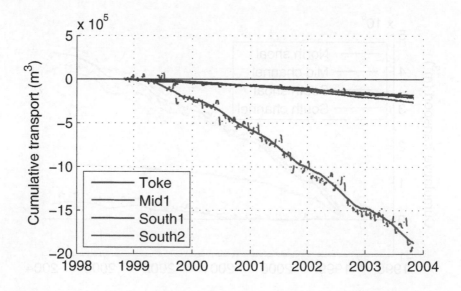

Figure 5.34 – *Time series of cumulative sediment transport through Inner Estuary control sections computed during brute-force simulation with all inputs simplified bf7 (solid lines) and an accelerated simulation with identical inputs ac1 (dash-dot lines).*

Table 5.7 – Brute-force simulation results.

Control Section	Transport $bf1$ (m^3)	Transport error						$bma\text{-}bf7$ (%)	Transport bma (m^3)	Error $bma\text{-}bf1$ (%)
		$bf2$ (%)	$bf3$ (%)	$bf4$ (%)	$bf5$ (%)	$bf6$ (%)	$bf7$ (%)			
Outer Coast										
Grayland	3.5×10^6	-5	-0	-22	-32	-5	-18	2	2.9×10^6	-17
Leadbetter	4.1×10^5	-82	6	-656	-241	26	-15	-36	2.2×10^5	-45
Long Beach	9.2×10^6	-1	0	-20	-19	-0	-8	-2	8.2×10^6	-11
Entrance Shoal										
North_shoal	1.5×10^6	3	0	10	2	-4	-1	2	1.5×10^6	1
Mid_channel	5.6×10^5	-29	0	-56	-19	23	-28	47	5.9×10^5	5
South_shoal	4.5×10^6	-3	1	-1	0	2	-0	4	4.7×10^6	4
South_channel	4.2×10^6	-6	1	1	1	-2	-6	4	4.1×10^6	-2
Entrance Channel										
Main_channel	-1.9×10^7	2	-0	2	1	0	2	0	-1.9×10^7	-6
Nahcotta	-6.1×10^6	1	0	-0	3	2	3	-1	-6.0×10^6	-4
Entrance_total	-8.4×10^6	-2	1	1	2	2	-0	7	-7.8×10^6	7
Inner Estuary										
Toke	4.8×10^4	37	-9	25	10	-495	-474	-6	-1.9×10^5	-498
Mid1	-1.7×10^6	1	0	-1	5	-15	-12	-4	-1.9×10^6	-16
South1	-2.1×10^5	-2	0	-1	2	-2	-2	-5	-2.2×10^5	-7
South2	-2.6×10^5	-0	0	0	2	-3	-1	-1	-2.6×10^5	-3

5.3.8   Sensitivity to wave input reduction

Comparison of the results of the brute-force simulations identified wave input reduction as the largest single contributor to error in the baseline accelerated simulation *ac1*. As the method of wave input reduction used for the baseline accelerated simulation was fairly arbitrary and unlikely to be optimum it seems worthwhile investigating the sensitivity of the results to alternative wave input reduction techniques. Several arbitrary decisions were made in the process of simplifying the wave climate. These include: the random order of the wave classes within each season, the number of wave classes within each season, the need to preserve the chronology of the individual years 1998 to 2003, and the need to retain the distinction between summer and winter seasons. In addition, the entire method of manually arriving at the selected number of wave classes based on the assumption that the morphological impact of waves is proportional to $H_s^{2.5}$ is highly questionable. A series of accelerated test simulations were therefore performed, the first six of which were designed to test the above assumptions. Table 5.4 records the differentiating characteristics of each of the accelerated simulations. These differences are elaborated below:

wr1 This simulation is identical to *ac1* except that a different random order is assigned to the wave classes within each season.

wr2 This simulation abandons the chronology of the individual years. All winter seasons are modelled identically, and similarly for all summer seasons. The intention behind this simulation is that the chronology of yearly wave and wind climates is not known in advance, so this is the best that could be achieved for making predictions of future morphological change.

wr3 This simulation abandons both the chronology of the individual years, and the seasonality of winter and summer. The 19 original wave classes (12 from winter and 7 from summer) are randomly reordered within the year. All years are also identical. This simulation was included as an example of a "very crude" wave simplification.

wr4 This simulation reduces the number of wave classes used to 6 classes each winter plus 4 classes each summer. The simulation was intended to be another "very crude" wave simplification.

wr5 This simulation uses a different simplified wave climate selected by use of the "Opti" elimination tool. In this case Opti was run on a set of 50 short simulation results, each running for one morphological tide. Each simulation contained a wave class and a mean wind with an equivalent wind stress to the winds occurring during that wave class *over the duration of the entire 5-year period*

wr6 This simulation uses another simplified wave climate arrived at using the "Opti" tool. In this case, Opti was run on a set of 405 short simulations results. In this case each simulation contained a wave class and the wind with the equivalent wind stress occurring during that wave class *over the duration of each separate season.* This approach allows Opti to better select wave+wind combinations better adjusted to each individual season.

Errors in residual sediment transport through each of the control sections for each of the wave reduction simulations are presented in Table 5.8 and the BSS for each of the model regions are presented in Table 5.9. Key results are:

1. Comparing *wr1* with *ac1*: Random re-ordering of the wave classes within the seasons makes a slight difference to the net longshore transport through the Outer Coast Grayland control section. Cumulative transports *during* each season diverge because of the altered order of wave classes however, if the system was behaving in an entirely linear manner, the cumulative transport should be identical at the end of each complete season. As indicated in Figure 5.35 this is generally the case, however a slight divergence of transport can be seen through the Grayland control section from the end of the winter of 2001-2002. A similar slight divergence is observed in transports through the North Shoal control section (not shown). Somewhat surprisingly, the BSS for the Outer Coast region is significantly reduced (from 0.86 to 0.79) by the change in wave class order. The BSS of the Entrance Shoal is not significantly changed. This result shows a sensitivity of the deeper portions of the rapidly changing outer delta to the chronology of wave events *within a season.*

2. Comparing *wr2* with *ac1*: Removing the chronology of the year by year wave climate significantly increases the error in residual transport through all Outer Coast and Entrance control sections. Part of this deterioration may be due to a poor selection of representative wave classes for each of the two remaining seasons as the increase in error through the Outer Coast sections should not occur if the wave input reduction procedure was performed well, although it is possible that this degradation could be due to the loss of the varying year by year correlation between wind and waves.

3. Comparing *wr3* with *wr2*: Removing seasonality in addition to removing annual chronology from the wave forcing introduces little extra error. However the resulting BSS for the Outer Coast (0.67) and Entrance Shoal (0.86) are the worst produced by any simulation.

4. Comparing *wr4* with *ac1*: Reducing the number of wave classes from 19 to 10 almost halves the duration of the simulation, but has little impact on the accuracy of the result. Longshore transport rates for the 10 wave class simulation are among the most accurate of any of the simulations.

5. Comparing *wr5* with *ac1*: Introduction of a new wave simplification based on the Opti technique using overall average wind stresses somewhat increases the error in transport through Outer Coast and Entrance control sections. Using average winds, the Opti technique has performed somewhat worse than the manual wave input reduction method, this is despite the Opti wave simplification including more wave classes (15 in winter and 10 in summer).

6. Comparing *wr6* with *ac1*: Using a new wave simplification based on Opti
   trained on simulations using separate average wind stresses for each wave
   class for each season performs well. This simulation produces the highest BSS
   (0.94) for the Outer Coast region of any accelerated simulation. Surprisingly,
   the residual transports through the Outer Coast control sections are not
   nearly the best of the accelerated simulations. The excellent BSS must be
   due to accurate prediction of the locations of deposition at the outer edge of
   the ebb-tidal delta.

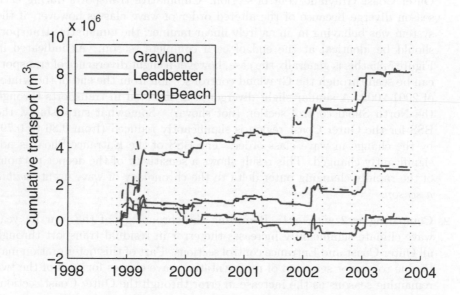

Figure 5.35 *– Time series of cumulative sediment transport through Outer
Coast control sections computed during accelerated simulation with original simplified wave input series ac1 (solid lines) and an accelerated simulation with the
order of wave classes within each season randomly altered wr1 (dash-dot lines).
Vertical red dashed lines indicate the limits of each season.*

Table 5.8 – *Accelerated simulation results.*

Control Section	Transport $bf1$ (m^3)	$ac1$	Transport error (%)										
			$wr1$	$wr2$	$wr3$	$wr4$	$wr5$	$wr6$	$tr1$	$tr2$	$tr3$	$tr4$	$tr5$
Outer Coast													
Grayland	3.5×10^6	−17	−12	−40	−43	−0	−59	−37	−17	−23	−18	−16	−25
Leadbetter	4.1×10^5	−45	−45	−349	−363	274	−452	−401	−29	−95	105	−8	34
Long Beach	9.2×10^6	−11	−10	−43	−43	6	−16	−14	−11	−11	−11	−11	−10
Entrance Shoal													
North_shoal	1.5×10^6	1	9	17	19	15	12	7	3	−11	39	10	15
Mid_channel	5.6×10^5	5	14	−134	−156	−51	−38	−9	25	30	164	13	162
South_shoal	4.5×10^6	4	4	−6	−6	4	−2	1	8	−4	36	11	20
South_channel	4.2×10^6	−2	−4	−7	−9	1	−8	−8	0	−7	22	3	15
Entrance Channel													
Main_channel	-1.9×10^7	2	3	10	12	2	5	3	5	20	13	−0	34
Nahcotta	-6.1×10^6	2	1	3	5	−3	2	2	9	32	27	−1	66
Entrance_total	-8.4×10^6	7	9	10	12	7	5	5	18	41	79	9	110
Inner Estuary													
Toke	4.8×10^4	−498	−492	−475	−476	−562	−495	−487	−329	−496	1590	−234	480
Mid1	-1.7×10^6	−16	−17	−19	−14	−25	−18	−18	−2	18	96	−10	103
South1	-2.1×10^5	−7	−7	−9	−6	−11	−8	−8	−4	27	−4	−15	38
South2	-2.6×10^5	−3	−3	−2	−3	−5	−3	−3	1	30	2	−11	38

Table 5.9 – Accelerated simulation skill scores.

Model												
Region	Brier Skill Score											
	ac1	wr1	wr2	wr3	wr4	wr5	wr6	tr1	tr2	tr3	tr4	tr5
Outer coast	0.86	0.79	0.67	0.67	0.88	0.81	0.94	0.87	0.86	0.90	0.83	0.75
Entrance shoals	0.97	0.96	0.87	0.86	0.95	0.95	0.97	0.95	0.95	0.72	0.92	0.81
Entrance channels	0.99	0.97	0.96	0.95	0.99	0.98	0.98	0.98	0.94	0.85	0.98	0.89
Inner estuary	0.98	0.98	0.98	0.98	0.98	0.98	0.98	0.99	0.90	0.87	0.99	0.91
Overall	0.96	0.94	0.90	0.89	0.96	0.94	0.97	0.96	0.93	0.83	0.94	0.85

5.3.9   Sensitivity to morphological tide

The brute-force simulations confirmed that the morphological tide selected for the accelerated simulations of Willapa Bay generally does an excellent job of reproducing the residual sediment transports through control sections. The only significant ($> 10\%$) errors introduced by use of the morphological tide were through the Leadbetter, Mid Channel, Toke, and Mid1 control sections (refer to column *bf6* in Table 5.7). Residual transports through the Leadbetter and Mid Channel control sections are very sensitive to small changes in forcing and in both of these sections the error introduced by the morphological tide is insignificant compared to errors introduced by other input reductions. At the Toke and Mid1 control sections the morphological tide is the main contributor to the error in the residual transport through these sections, and as such it is interesting to investigate them a little further.

One hypothesis worth considering is that as both the Toke and Mid1 locations inside the estuary are in channels with little residual flow then, according to the analysis performed in Section 5.2.2, residual transport should be insensitive to the conservation of total energy in the morphological tide, but may be detrimentally affected by the multiplication factor applied to the $(M_2 + C_1)$ constituents. The baseline accelerated simulation *ac1* uses a morphological tide of $1.08 \times (M_2 + C_1)$, however five more accelerated simulations were performed as follows:

tr1 This simulation is identical to simulation *ac1* except that it uses a morphological tide of $1.09 \times M_2 + C_1$.

tr2 This simulation is identical to simulation *ac1* except that it uses a morphological tide of $1.00 \times (M_2 + C_1)$.

tr3 This simulation is identical to simulation *ac1* except that it uses a morphological tide of $1.20 \times M_2$.

tr4 This simulation is identical to simulation *ac1* except that it uses a morphological tide of $1.12 \times M_2 + C_1$.

tr5 This simulation is identical to simulation *ac1* except that it uses a morphological tide of $1.08 \times M_2$.

Simulations *tr1* and *tr3* were designed to each have the same total energy as the natural tide. Simulation *tr2* applies no factor to the "pure" $M_2 + C_1$ tide, simulation *tr4* has the same energy as the tide $1.08 \times (M_2 + C_1)$ used for the baseline accelerated simulation *ac1*, and simulation *tr5* uses the original morphological tide used in previous studies and the simulations reported in Chapter 4.

The resulting errors in residual transports through the control sections are presented in Table 5.8 and the corresponding BSS in Table 5.9. Key results are:

1. Comparing simulation *tr1* with simulation *ac1*: Use of the $1.09 \times M_2 + C_1$ morphological tide reduces the error in the cumulative sediment transport through the Toke control section by about 35% and completely eliminates the error in the transport through the Mid1 control section (see Figure 5.36. The overall BSS for the modified morphological tide are not significantly

different, this just emphasises the fact that the residual transport errors well inside the estuary are very small and produce insignificant morphological change in the context of change in the estuary entrance.

2. Use of the $1.09 \times M_2 + C_1$ morphological tide increases error in the Entrance Shoal and Entrance Channel regions. This is caused by the slight reduction in total energy of the morphological tide interacting with the strong residual currents through these control sections as can be seen when comparing simulation *tr4* with simulation *tr1*. Simulation *tr4* has a morphological tide of $1.12 \times M_2 + C_1$ which has the same total energy as the tide used in the baseline accelerated simulation *ac1*. This slight increase in energy reduces the error in the transports in the main entrance channels to near zero, however the error in transports over the entrance shoals is increased, as is the error through the inner estuary control sections, except for section Toke (Table 5.8). Clearly the "best" morphological tide will never perfectly match all residual transports and the optimum choice of the amplification factor will depend on which areas of the model are more critical to the modelling objectives. The selection of amplification factor could easily be achieved on the basis of a BSS computed over the region of greatest importance. In this case, simulation *tr1* has better BSS than *tr4* for the Outer Coast and Entrance Shoals regions and therefore would be more suitable for the objectives of the present study.

3. Comparing simulation *tr2* with simulation *ac1*: Using a tide of just $M_2 + C_1$ which, according to the theory laid out in Section 5.2.2, should provide an accurate residual transport in areas of insignificant mean flow returned nearly identical residual transport through the Toke section to the $1.08 \times (M_2 + C_1)$ simulation. This is an unexpected result and indicates that some other factor is at play in determining the tidal influence on transports through the Toke control section. It is also clear that although an entire control section may have close to zero residual flow across it it is still quite possible for individual grid cells, or subsections of the control section to experience significant residual flows due to horizontal tidal asymmetries. This is clearly the case for the Toke control section.

4. Comparing simulation *tr3* with simulation *ac1*: Neglecting the C1 constituent altogether and increasing the energy in the M_2 component to compensate – something akin to the "standard" method of determining a morphological tide – produces a very poor result, with significant positive (flood-directed) errors through most control sections. The original morphological tide simulation is repeated in simulation *tr5*

5. Simulation *tr5* was performed to demonstrate the improvement gained by using the new morphological tide consisting of $1.09 \times M_2 + C_1$ compared to the original morphological tide of $1.08 \times M_2$ derived using the previous "standard" methodology, as used in Chapter 4. Comparing simulation *tr5* with simulation *tr1* it is clear that the new morphological tide significantly improves residual transport through virtually all control sections (Table 5.8).

This is reflected in the BSS with the overall BSS improving from 0.85 to 0.96 with the introduction of the new morphological tide. The BSS for all model regions are substantially improved.

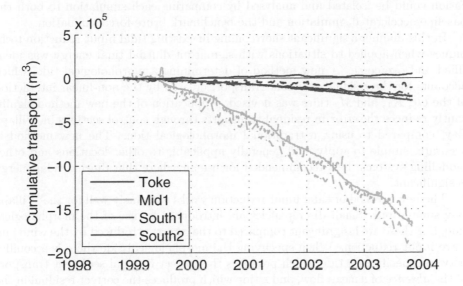

Figure 5.36 *– Time series of cumulative sediment transport through Inner Estuary control sections computed during accelerated simulation with baseline morphological tide ac1 (solid lines), an accelerated simulation with a morphological tide of* $1.09 \times M_2 + C_1$ *wt1 (dash-dot lines), and the benchmark brute-force simulation bf1.*

5.4   Conclusions

This chapter set out to analyse input reduction and morphological acceleration techniques suitable for use with the new online morphological model developed in Chapter 2. Objectives were to determine what modifications to existing techniques were required to work with the new model, and then to quantify the magnitude of the errors introduced by the use of the techniques when applied to a typical medium-term morphological modelling problem. These objectives were achieved by initially performing a series of brute-force simulations, each of which contained one simplified forcing process. The results of these simulations were compared with a benchmark brute-force simulation using the full set of forcing processes. This allowed the error introduced by the simplification of each forcing process to be isolated and analysed.

A number of simulations accelerated using the new "variable morfac" morphological acceleration technique were then performed. The first simulation included exactly the same simplified forcing processes as tested using the brute-force simulations. Comparing this simulation against the brute-force simulations permitted the error introduced by the morphological acceleration technique to be isolated and analysed. A short series of accelerated simulations were then performed to test sensitivity and possible improvements to the wave and tidal input reduction techniques applied to the initial series of simulations. The impact of each modification could be isolated and analysed by comparing each simulation to both the baseline accelerated simulation and the benchmark brute-force simulation.

In this manner a significant shortcoming in existing tidal input reduction techniques when applied to situations with significant diurnal tidal energy was identified and overcome. A new method of determining a morphological tide, which accounts for the residual sediment transport created by the non-linear interaction of the O_1, K_1, and M_2 tides was devised. Application of the new method significantly reduces the error in residual transports through control sections in Willapa Bay, compared to using a traditional morphological tide. The new method is accurate, simple to apply, and generally applicable to other locations and other modelling systems. It is recommended for use anywhere that diurnal tidal energy is significant.

The new method of tidal input reduction works extremely well for the Willapa Bay simulations. Small discrepancies are introduced by use of the morphological tide, but these are insignificant compared to the errors introduced by the wind and wave input reduction. When specifying the morphological tide there is a conflict between specifying a tide which produces the correct residual sediment transport in the *absence* of a mean flow, and a tide which produces the correct residual in the *presence* of a mean flow. Tests confirm that the suggested method of specifying a morphological tide consisting of a $factor \times M_2 + C_1$ minimises the error introduced. Significant error was observed in the (small) residual sediment transport produced by the morphological tide in an area of complex divergence of residual transports near the Toke control section, well inside Willapa Bay. Investigating the reasons for the error in this particular location could be a fruitful area for further research.

In applying the model and techniques to 5-year morphological simulations of Willapa Bay it was determined that the method used to simplify the wind and wave forcing processes contributes by far the most error to the accelerated simulations. Two methods of wind and wave input reduction were tested. The straightforward, manual, method of selecting representative wave classes could perform reasonably well, but this study could not obtain consistency in the results achieved. Simulations using fewer wave classes could out-perform simulations with a greater number. This should not occur if the method of selecting wave classes was working well. Only one of the simulations did a good job of accurately capturing the longshore sediment transport along the outer coast. This should be a basic requirement for any wave input reduction technique. The alternate "Opti" method of wave input reduction tested did an excellent job of capturing the morphological change of the benchmark simulation, but only when it was "trained" on a series of more than 400 base simulations, each one containing a unique combination of wave and wind conditions. This seems an extreme level of effort to achieve a satisfactory simplified wave plus wind climate. Using the Opti approach with a simpler training set of 50 base simulations, using mean wind stresses, achieved similar results as mediocre manual selection of the wave classes. Other, more refined, manual wave input reduction techniques have been discussed in the literature, but were not tested here. It is highly probable that an established technique could achieve better results than obtained by this study. The main conclusion in this regard is that existing techniques in the literature should be investigated in more detail, with proper analysis of the errors introduced, because this is clearly the "weakest link" when producing accelerated morphological simulations of coastal environments such as Willapa Bay.

In Willapa Bay both wind and waves play an important role in determining residual sediment transport. The occurrence of wind and wave events are partially correlated and it is essential that this correlation is captured by the input reduction methods. The relatively simple method of computing the mean wind *stress* vector occurring during each wave class, and then applying a wind producing an equivalent stress during the simulation of each wave class worked reasonably well, introducing less error into the model than wave input reduction. One key factor to be aware of is that wind-driven currents take some time to develop, especially in deeper water. This means that wind-driven currents must be given some time to develop, especially in a morphologically accelerated model. This effect is perhaps not as critical as it might otherwise be, as wind-driven currents develop relatively rapidly in shallow water where the bulk of sediment transport takes place.

In Willapa Bay, neither river discharge, nor low frequency water level fluctuations contribute greatly to residual sediment transport patterns. Using very simple input reduction techniques (mean values per season) for these forcing processes introduced no significant error into the simulation. If the processes had a greater impact then employing a similar procedure as used for wind would probably have been sufficient.

The total error introduced by simplifying all forcing processes was close, but not quite equal to, the sum of the individual errors. Some degree of non-linearity can be expected when processes with an important role in driving residual sediment transport are partially correlated, as simplifying these processes separately destroys the correlation between them. This was the case for wind and waves at Willapa Bay.

The chronology of the "random" forcing by waves was found to play a significant role in Willapa Bay. Abandoning the known annual sequence of individual years resulted in a significant degradation of the model results. This highlights the need for ensembles of simulations to be performed for times when the annual chronology of forcing is unknown (such as for future predictions).

The "variable morfac" method of morphological acceleration used at Willapa Bay is confirmed to be a simple and robust method of accelerating morphological developments. It is relatively simple method to construct the required input files and to check the resulting simulation has run correctly. The morfac method is extremely flexible and can easily be used to accommodate a wide array of modelling problems. The error introduced into the Willapa Bay simulations by the use of the variable morfac approach was negligible and in the same order as the error caused by the use of the morphological tide. The morphological acceleration approach was very efficient, allowing the accelerated simulations to be completed in less than 10% of the time taken for the brute-force simulations. Further improvements are possible using further enhancements of the variable morfac approach, such as the "parallel online morfac" approach described by Roelvink (2006). One trap exists with the morfac approach, where a loss of continuity of sediment mass will occur if the morfac value is changed while sediment is in suspension. This problem can be minimised by carefully choosing the times at which morfac is changed. This potential problem also applies to the start and end of a simulation using a constant morfac value.

The approach used in this study, contrasting the results of multiple brute-force simulations with different simplified forcing processes, worked well for isolating the errors introduced by each input reduction. Similarly, contrasting these results with an otherwise identical accelerated simulation allowed the successful isolation of the errors caused by that technique.

Quantifying the errors by way of cumulative transports through control sections and the use of Brier Skill Scores (BSS) computed for discrete regions of the model provided an excellent understanding of the impact of the various changes on both time histories of sediment transport and resultant morphological change.

The input reduction and morphological acceleration techniques described here are clearly not the weakest link in the medium-term morphological modelling of Willapa Bay. The total error introduced by input reduction and morphological acceleration (including the less than optimal wave input reduction used in this study) resulted in an overall BSS of 0.96 relative to a baseline of zero morphological change. To quote an alternative statistic, using the input reduction and morphological acceleration techniques described in this chapter introduced a mean absolute error in computed morphological change of 0.13m out of an observed mean absolute signal of 1.9m. This error ($< 7\%$ of the signal) is a small price to pay for accelerating a simulation by at least a factor of ten. To put this in context; despite showing many of the qualitative features of the observed morphological change patterns, the benchmark brute-force simulation achieved an overall BSS of -0.47 and a mean absolute error of 2.4m relative to observed morphological changes. Clearly there is plenty of scope to improve the predictive skill of the model.

It is therefore clear that, if the underlying wave and hydrodynamic models can be considered well validated (Chapter 4), then the greatest errors in the morphological model must be caused by shortcomings in the physical process descriptions included in, or missing from, the (residual) sediment transport model. This study concludes that, at least for estuaries such as Willapa Bay, development and testing of improved sediment transport formulations can confidently proceed using accelerated morphological models driven by simplified forcing processes.

The acceleration techniques used in this chapter achieved an acceleration in excess of 10 times the speed of the brute-force simulation, however one simulation showed that similar results could be achieved using fewer, carefully selected, wave classes – achieving an acceleration factor of approximately 20. A further enhancement to the variable morfac acceleration technique described in this chapter – the so-called "parallel online" approach – would yields greater acceleration by running multiple morphological tides with associated representative wave conditions in parallel. This approach requires no additional assumptions and could achieve acceleration factors of 100 or more, as long as sufficient computer hardware is available. Assuming that the basic hydrodynamic and sediment transport models are efficiently designed and run at a speed in the order of 10 to 50 times real time then the resulting accelerated morphological simulations would run at somewhere between 1000 and 5000 × real time. This implies that $20 - 100$ years of morphological development can be simulated in a week of computer time, which is a reasonable duration for performing a limited sensitivity analysis. This type of simulation should be ample for most "engineering" applications of process-based morphological models. The overarching conclusion of this chapter is therefore that the morphological acceleration techniques discussed in this chapter are sufficiently accurate, robust and efficient for most "medium-term" problems and that the most pressing challenges in morphological modelling lie in improving the basic physics encapsulated in the sediment transport models.

Chapter 6

Conclusions

The central aim of this research is to develop and test improved methods and modelling approaches for the analysis and prediction of coastal morphology on spatial scales of kilometres and time scales of several years. This objective has clearly been achieved as the modelling approach and several of the methods developed and validated in the course of the work described here are already in widespread use by engineers and researchers around the world.

In order to achieve this overarching objective, four specific objectives were set. Conclusions reached on the way to these objectives are as follows:

6.1   Developing a 3D Morphological Model

Objective: To develop a new generally applicable coastal morphological model capable of representing the dominant coastal sediment transport processes in a fully three-dimensional manner and to simulate the resulting morphological change occurring over a period of several years. The model should also be simpler to apply and more robust than existing coastal morphological models.

A new coastal morphological model has been developed by adding sediment transport and morphological updating formulations to the existing Delft3D-FLOW hydrodynamic model. Following testing and validation the model has been adopted by Delft Hydraulics as its new standard morphological model for coastal (and some fresh water) applications. This was an efficient method to develop a new model as it built on the robustness and power of an existing advanced three-dimensional hydrodynamic flow solver which had been widely tested and shown to reproduce a wide array of coastal processes.

The new model differs from previous models in two main respects: First, it is fully three-dimensional and second, the sediment transport and morphological updating are tightly coupled with the hydrodynamics. These two aspects give the new modelling approach several advantages over previous approaches:

1. Being three-dimensional, the model can explicitly include the complexities of three-dimensional hydrodynamics of the coastal zone, to the extent that these are understood and implemented in the three-dimensional flow solver. These can include the three-dimensional effects of waves on hydrodynamics and sediment transport and the effects of density stratification and density-driven currents.

2. Solving the sediment transport equations in three dimensions also explicitly accounts for adaptation of non-equilibrium sediment transport rates. Suspended sediment transport is not assumed to be in local equilibrium with bed shear stresses. This contributes to natural, smooth, model results.

3. The tight integration of the new formulations into the Delft3D-FLOW model means that it is relatively simple for users to construct morphological models which require repeated updating of wave, tide, and sediment transport patterns in two or three dimensions. This increases the efficiency of constructing new models and lowers the barrier to using a powerful process-based model for analysing coastal morphological problems.

4. Tight coupling of the sediment transport formulations with the hydrodynamic computations allows the straightforward implementation of sediment-related effects on the hydrodynamics (such as turbulence damping and density currents). It also allows robust treatment of drying and flooding grid cells, non-erodible layers, and the implementation of a simple, but effective, dry-bank erosion formulation.

5. The very frequent feedback of morphological change to the hydrodynamic calculations helps keep the model extremely stable without introducing a lot of numerical diffusion. A robust and stable model is essential for performing long-duration morphological simulations.

6. A new method of morphological acceleration (morfac) has been introduced and has been shown to be robust and conceptually simple to apply. The method has many of the advantages of previous acceleration methods and few of the disadvantages.

The interaction between the hydrodynamic model and the wave model has also been improved to more realistically model both the three-dimensional effects of wave forces on coastal hydrodynamics and of three-dimensional current structures on spectral wave propagation. In both cases relatively simple "engineering" approximations to these complex interactions have been implemented and doubtless they can be further improved in the future. However, testing indicates that the approaches implemented are adequate for many coastal applications.

Two-dimensional (depth averaged) formulations have also been implemented in the model in a manner that mimics the three-dimensional implementation. Depth averaged modelling requires more assumptions and approximations than modelling in three-dimensions, however depth averaged models run substantially faster than their three-dimensional counterparts and may be appropriate in some instances. As time progresses, ongoing increases in computing power should lead to a reduced need to perform depth averaged modelling.

6.2   Validation of Process Models

Objective: To validate the individual process models contained within this new model against available theories, laboratory, and field data.

The new model has been tested against a wide range of validation cases. These included theoretical results such as the development of sediment transport in an initially clear flow, establishment of an equilibrium longitudinal bed slope from an initially horizontal bed, and simple settling of sediment from suspension. The model has also been validated against laboratory datasets under controlled conditions. The model can well reproduce laboratory experiments of a trench migrating downstream in a flume, the development of bars and channels in a curved flume with spiralling flow, and observed sediment concentration profiles under the action of waves and various speeds of following current. The new model also reproduced, or improved upon, previous model results for the hypothetical, but previously documented, cases of a sediment hump deforming under unidirectional flow and of the development of a tombolo behind an emergent shore-parallel breakwater.

In all cases the model has run reliably and produced smooth, realistic results. Individual processes have been easy to calibrate to match measurements although, for the test simulations performed during this research, little variation from the default settings has been required. This builds confidence in the individual process models' ability to reproduce the most important physical processes present in the coastal zone.

Application of the model to more complex real-life measurements of morphological change in the coastal zone has not been so straightforward however. Application to situations where both tidal currents and waves are important drivers of sediment transport and morphological change reveals that it is much more difficult to achieve reliable results. The model shows sensitivity to the choice of a number of parameterisations, especially to the choice of bed roughness formulation. The choice of Chezy, Manning, or constant K_s bed roughness formulations can lead to widely varying predicted morphological development, even in the simple case of a unidirectional current without waves. When the temporal variation in bed roughness known to occur in tidal regimes, but not captured in the model, is considered - along with the impact of wave orbital motions - it is clear that this is one critical area of the model formulations that is not yet sufficiently well described.

Similarly, the morphological behaviour of tidal channels in the model indicates that important processes which act on the side slopes of channels and largely dictate their cross-sectional shape may be missing from the present model formulations. The approach of vastly over-stating the down-slope transport of bed-load material used in the Willapa Bay application is not recommended as a long-term solution to this problem. This is clearly an area in which further research is required.

6.3   Application of the Model

Objective: To validate the complete morphological model in a complex coastal environment with a well-described morphological change signal and well known forcing conditions for a period of several years.

Application of the model and morphological acceleration techniques to the morphodynamics of a complex and dynamic ebb-tidal delta demonstrated the abilities and shortcomings of the present state of the art in process-based coastal morphological models. The model predictions of morphodynamic behaviour on the timescale of years show some qualitative skill - most of the observed general patterns are reproduced - but the magnitude and/or precise location of the changes are not predicted well by the model. Use of an objective Brier Skill Score to measure model performance results in a negative skill score, or a model skill worse than simply predicting that no morphological change occurs.

This lack of skill cannot be attributed to the input reduction and morphological acceleration techniques employed (see following section), but rather to either the initial conditions (i.e. bathymetry) or elementary physics contained in the model – most probably the latter as in simulations of recent morphological developments the initial bathymetry is well known.

Investigation of unexpected morphological feedback mechanisms and testing the morphological model sensitivity to changed model settings required a series of many, many morphological simulations. This is precisely why the input reduction and morphological acceleration techniques discussed in this thesis are essential - without them long-term morphological simulations are so time-consuming that a reasonable investigation of model sensitivity is simply not feasible. Accelerated simulations should be used for these sensitivity investigations, at a minimum.

Use of a process-based model, as part of a broader investigation into coastal erosion problems, was extremely beneficial. The qualitative skill of the morphological model gives confidence that the computed (residual) sediment transport patterns broadly represent the main sediment pathways. Use of the model to isolate the effects of individual forcing processes and to test the sensitivity of the result to changes in initial conditions and/or forcing contributed greatly to understanding the likely cause(s) of observed morphological behaviour. Careful interpretation of model results in the context of other observations and sensitivity analyses remains essential.

The collection of field data in coastal environments is not without potential pitfalls. Post processing of field data is usually required and errors may be introduced either in instrument configuration or at this latter stage. Comparison of field measurements with the results of a good process model (even an uncalibrated one) provides an excellent check on field data collection and processing. Neither the model nor the data should be blindly trusted.

Brute-force morphological model simulations also have an important role to play in any investigation. Temporal variation in forcing processes will always be important at some scale and using brute-force simulations to benchmark accelerated simulations is highly recommended. Visual inspection of brute-force model time series output is also valuable for understanding the role of individual physical processes.

6.4   Morphological Acceleration Techniques

Objective: To isolate and quantify the errors introduced by use of the morphological acceleration methods used with the new model and to compare the significance of these errors with other errors present in the modelling process.

To undertake morphological simulations with durations greater than just a few weeks using a process-based morphological model, some form of morphological acceleration techniques will be required. In coastal environments, use of the morfac approach implemented in the new model requires that forcing processes be reduced to a limited number of representative conditions. This will typically require both tidal and wave input reduction. Other processes may also have to be reduced to representative values, depending on the location in question.

Input reduction must be performed carefully, considering both the mean effects of forcing processes on sediment transport and also the non-linear interactions that may occur between correlated forcing processes (e.g. wind and waves). Simple methods exist to capture the most important of these interactions.

Investigation of errors introduced by a previous "standard" method of tidal input reduction identified an important shortcoming of the method when applied to locations with significant diurnal tidal energy. This thesis confirms previous findings that a significant residual sediment transport will result from the interaction of the M_2, O_1, and K_1 tidal constituents and that shallow water overtides are not required to produce residual sediment transport when significant diurnal tidal energy is present. A method has been devised to capture this extra residual sediment transport in a new simple morphological tide which consists of just two harmonics and is periodic at twice the M_2 period. Tests confirm that this new morphological tide performs considerably better than the previous tide in an estuary subject to mixed diurnal and semi-diurnal tidal forcing.

Selection of the ideal morphological tide depends on the exponent used in the sediment transport formula and the level of mean (residual, non-tidal) flow. As both of these factors vary in space (the exponent will effectively vary depending on the degree of wave stirring experienced) no morphological tide will be perfectly representative at all locations in a model domain. This implies that some compromise morphological tide will need to be selected based on the modelling objectives of the study in question. Results achieved during this research suggest that any errors introduced are likely to be insignificant compared to other errors in the input reduction and physical process descriptions in the model.

The methods of wave input reduction tested during this study were satisfactory, but could doubtless be improved. It should be possible to select a representative set of wave classes which closely approximates the longshore transport produced by the full wave climate, while still providing the correct level of sediment stirring and wave-driven currents in other areas of the model. This is critical to minimising the errors introduced into an accelerated morphological simulation. Reproducing the degree of the correlation between waves and other "random" forcing processes, such as wind, is also critical. Further work may also be required in this area once an improved method of wave input reduction is found.

The "variable morfac" approach to accelerating a morphological simulation in the presence of complex wave and wind climates worked well and introduced a negligible amount of error into the simulations. Care needs to be taken when *changing* morfac values during any accelerated morphological simulation to avoid introducing sediment mass conservation errors. Any errors can be kept to manageable levels by ensuring that suspended sediment concentrations are similar at times when non-zero morfac values are started, stopped, or changed.

The variable morfac acceleration approach used in this study achieved a net acceleration in excess of 10 times the speed of the brute-force simulation. Tests indicate that by reducing the number of wave classes used this acceleration could be increased to a factor of approximately 20 without degrading the result. If further acceleration is required then using morfac but computing the residual sediment transport due to multiple wave and tidal conditions in parallel (the so-called "parallel online" approach) has been shown by other authors to produce results at least as accurate as the "variable morfac" approach and to achieve acceleration factors of 100 or more. These levels of acceleration should allow the simulation of several decades of morphological development in less than 1 week of computation time. This seems sufficient for most "medium-term" morphological modelling requirements.

When wave effects can be neglected and morphological developments take place at a slower rate higher morphological acceleration factors can be used as the frequent feedback from morphology to hydrodynamics employed in the "online" approach keeps the model extremely stable. Several authors have now demonstrated the use of high morfac values to successfully model several *thousand* years of coastal morphological development under simple tidal forcing.

Quantification of model performance by use of skill scores and other quantitative methods is essential for two- or three-dimensional model validation. Visual comparison of model and measurement data sets is useful for identifying qualitative patterns in the results but can be very misleading if quantitative model accuracy is required. The use of separate Brier Skill Scores for different model regions, as used in this study, was extremely helpful in this regard.

6.5 Concluding Remarks

In summary, the morphological modelling approach described in this thesis makes important advances over previous models and methods. It is relatively simple to set up and apply to applications from laboratory to field scales. It contains state of the art descriptions of many of the physical processes occurring in the coastal zone. It includes a simple, but effective, technique for extending the effective duration of simulations to morphological scales of interest. The model has already been applied, by several researchers, to a wide array of coastal problems and has been shown to be a useful tool for conducting research and tackling engineering questions.

The model is not without its faults however. The ability to simulate relatively long periods highlights the important role of the dynamic feedback that occurs in fully coupled morphological models and that further investigation is required to better understand this. A related challenge is to develop methods of "tuning" morphological models to observed morphological behaviour through learning which parameter settings in which physical processes play an important role in determining the computed morphological development. The tuning of systems of complex processes-based models is not intuitive. The feedback processes (from sediment transport to hydrodynamics via morphology) are poorly understood and are critical to long-term morphological models.

A number of morphologically important physical processes may still be missing from the model. Results presented in this thesis indicate that the choice of bottom roughness formulation can dramatically alter the computed morphological development. This is worrying as it is well known that the roughness formulations currently available fail to capture even a fraction of the real spatial and temporal complexity occurring in nature. Results also strongly indicate that (sub-grid) transport of sediment on the banks of unrestrained tidal channels may play an essential role in shaping the long-term morphology of those channels, yet this process is not adequately described in the present model. These are areas where ongoing research into the fundamental physics of sediment transport will play a critical role in improving the current state of the art models. The models themselves can assist in this process by being used as numerical laboratories in which to test new ideas and formulations. Morphological model formulations will therefore remain fairly "dynamic" for some time to come.

List of Symbols

Symbol	Meaning
a	Reference height for suspended sediment concentration (m)
c	Mass sediment concentration (kg/m^3)
c_a	Mass sediment concentration at reference height a (kg/m^3)
d	Depth to bed from reference datum (positive down) (m)
d_{50}	Median sediment diameter (m)
D_H, D_V	Horizontal and vertical diffusion coefficients (m^2/s)
D_*	Dimensionless sediment diameter (-)
f	Coriolis coefficient (inertial frequency) (s^{-1})
h	Water depth (m)
H_s	Significant wave height (m)
k	Turbulent kinetic energy (m^2/s^2)
P	Pressure (Pa)
S	Salinity (ppt)
S_b	Bed-load sediment transport (kg/m/s)
T	Dimensionless shear stress (-)
T_p	Peak wave period (s)
u, v, w	Eularian velocity components in Cartesian coordinates (m/s)
U, V	Generalised Lagrangian Mean (GLM) velocity components (m/s)
u_*	Bed shear velocity (m/s)
$\bar{U}, \bar{V}$	Depth-averaged GLM velocity components (m/s)
$\hat{U}_\delta$	Peak orbital velocity at the bed (m/s)
w_s	Sediment fall velocity (m/s)
z	Vertical Cartesian coordinate (m)
β	Ratio of sediment diffusion to fluid diffusion (-)
ε	Dissipation of turbulent kinetic energy (m^2/s^3)
κ	Von Karman's constant (=0.41)
ν_H, ν_V	Kinematic viscosity (m^2/s)
ρ	Local fluid density (inc. salinity, temperature and sediment)(kg/m^3)
ρ_0	Reference density of water (kg/m^3)
ρ_s	Density of solid sediment particles (kg/m^3)
σ	Vertical "sigma" coordinate (-)
τ_{bx}, τ_{by}	Bed shear stress components (N/m^2)
ω	Vertical velocity component in sigma coordinate system (s^{-1})
ζ	Water surface elevation above reference datum (m)

212 LIST OF SYMBOLS

Bibliography

Allan, J.C., and Komar, P.D., 2000. The wave climate of the eastern North Pacific: Long-term trends and El Niño/La Niña dependence. *Abst. Proceedings Fifth Southwest Washington Coastal Erosion Study Workshop*, pp. 129-130.

Andrews, R.S., 1965. *Modern sediments of Willapa Bay Washington: A coastal plain estuary.* Department of Oceanography, Univ. of Washington, Technical Report 118, 43p.

Andrews, D.G. and Mcintyre, M.E., 1978. An exact theory of non-linear waves on a Lagrangian-mean flow. *J. Fluid Mechanics, Vol. 89* (4), pp. 609-646.

Bagnold, R.A., 1966. *An Approach to the Sediment Transport Problem from General Physics.* Geological Survey Prof. Paper 422-I, US Geological Survey, DOI, USA.

Ballard, R.L., 1964. *Distribution of beach sediments near the Colombia River.* Department of Oceanography, Univ. of Washington, Technical Report 98, 82p.

Banas, N.S., Hickey, B.M., MacCready, P., Newton, J.A., 2004. Dynamics of Willapa Bay, Washington: A highly unsteady partially mixed estuary. *J. Physical Oceanography, 34*, pp. 2413-2427.

Bernardes, M.E.C., Davidson, M.A., Dyer, K.R., George, K.J., 2006. Towards medium-term (order of months) morphodynamic modelling of the Teign estuary, UK. *Ocean Dynamics, Springer, Volume 56*, pp. 186-197.

Bijker, E., 1971. Longshore transport computations. *J. Waterways, Harbours and Coastal Engineering Division, ASCE, Vol. 97*, pp. 687-701.

Blumberg, A.F. and Mellor, G.L., 1985. Modelling vertical and horizontal diffusivities with the sigma co-ordinate system. *Monthly Weather Review, Vol. 113*, No. 8.

Booij, N., Ris, R.C., and Holthuisen, L.H., 1999. A third-generation wave model for coastal regions, Part I: Model description and validation. *J. Geophysical Research, 104*, C4, 7649-7666.

Bos, K.J., Roelvink, J.A., and Dingemans, M.W., 1996. Modelling the impact of detached breakwaters on the coast. *Proc. 25th Internatinoal Conference on Coastal Engineering*, ASCE, New York, pp. 2022-2035.

Cañizares, R., Alfageme, S., and Irish, J. L., 2003. Modeling of morphological changes at Shinnecock Inlet, New York, USA. *Proc. Coastal Sediments 2003*, Florida, Paper IV-B-5.

Carver, R.E., 1971. *Procedures in Sedimentary Petrology*, New York, John Wiley and Sons, 653 pp.

Cayocca, F., 2001. Long-term morphological modeling of a tidal inlet: the Arcachon Basin, France. *Coastal Engineering 42*:115-142

Clifton, H.E., and Phillips, R.L., 1980. Lateral trends and vertical sequences in estuarine sediments, Willapa Bay, Washington, in Field, M.E., Bouma, A.H., Colburn, I.P., Douglas, R.G., and Ingle, J.C., eds., Quaternary depositional environments of the Pacific Coast. *Pacific Section SEPM Paleogeography Symposium No. 4*, p. 55-71.

Dastgheib, A., Wang, Z.B., Ronde, de J., and Roelvink, J.A., 2008. Modelling of mega-scale equilibrium condition of tidal basins in the western Dutch Wadden Sea using a process-based model. *Proceedings of PIANC-COPEDEC VII Conference 2008, Dubai.*

Dekker, S. and Jacobs, C.E.J., 2000. *Sediment Concentration Due To Irregular Waves and Currents*. Master's thesis, Delft University of Technology.

Dingemans, M.W., Radder, A.C. and de Vriend, H.J., 1987. Computation of the Driving Forces of Wave-Induced Currents. *Coastal Engineering, 11*, pp. 539-563.

Dingemans, M.W., 1997. *Wave propagation over uneven bottoms*. Advanced Series on Ocean Engineering 13. World Scientific, Singapore. ISBN 981-02-0427-2

Dingler, J.R., and Clifton, H.E., 1994. Barrier systems of California, Oregon, and Washington, in Davis, R.A., ed., *Geology of Holocene Barrier Islands*. Springer-Verlag, Berlin, pp. 115-165.

Eckart, C., 1958. Properties of water, Part II. The equation of state of water and sea water at low temperatures and pressures. *American Journal of Science, 256*, pp. 225-240.

Edwards, T.K., and Glysson G.D., 1999. Field Methods for Measurement of Fluvial Sediment. *U.S. Geological Survey Techniques of Water-Resources Investigations, Book 3, Chapter C2 - revision*, Reston, VA, 97 pp.

Elias, E., 2006. *Morphodynamics of Texel Inlet*. IOS Press, Amsterdam. ISBN 978-1-58603-676-8

FitzGerald, D.M., Kraus, N.C., and Hands, E.B., 2001. *Natural mechanisms of sediment bypassing at tidal inlets.* ERDC/CHL-IV-30, U.S. Army Engineer Research and Development Center, Vicksburg, MS. (http://chl.wes.army.mil/library/publications/cetn)

Folk, R.L., 1968. *Petrology of Sedimentary Rocks*. University of Texas Publication, Austin, 170 pp.

Forester, C.K., 1979. Higher Order Monotonic Convective Difference Schemes. *Journal of Computational Physics, Vol. 23*, 1-22.

Fredsoe, J., 1984. Turbulent boundary layer in wave-current interaction. *Journal of Hydraulic Engineering, ASCE, Vol. 110*, 1103-1120.

Gallappatti, R., 1983. A depth integrated model for suspended transport. *Communications on Hydraulics, Vol. 83-7*, Delft University of Technology, Delft, 114p.

Gessler, D., Hall B., Spasojevic M., Holly F., Pourtaheri H. and Raphelt N., 1999. Application of 3D mobile bed, hydrodynamic model. *Journal of Hydraulic Engineering*, July 1999, pp. 737-749.

Gelfenbaum, G., Kaminsky G.M., Sherwood C.R., and Peterson C., 1997. *Southwest Washington Coastal Erosion Work- shop Report, 1997*, U. S. Geological Survey Open-File Report 97-471, 102 pp.

Gelfenbaum, G., and Kaminsky G.M., 2000. *Southwest Washington Coastal Erosion Workshop Report 1999*, U. S. Geological Survey Open-File Report 00-439, 187 pp.

Gelfenbaum, G., and Kaminsky G.M., 2002. *Southwest Washington Coastal Erosion Workshop Report 2000*, U. S. Geological Survey Open-File Report 02-229, 308 pp.

Gelfenbaum, G., Roelvink, J.A., Meijs, M., and Ruggiero, P., 2003. Process-based morphological modeling of Grays Harbor inlet at decadal timescales. *Proceedings of the International Conference on Coastal Sediments 2003*. CD-ROM Published by World Scientific Publishing Corp. and East Meets West Productions, Corpus Christi, Texas, USA. ISBN 981-238-422-7.

Gibbs, A.E., Buijsman, M.C., and Sherwood, C.R., 2000. *Non-navigational Grid ded Bathymetry Data: The Washington-Oregon Coast, 1926-1998; A data release and description of methods*. USGS Open-File-Report OF 00-448.

Grasmeijer, B.T. and van Rijn, L.C., 1998. Breaker bar formation and migration. *Proceedings of the 26th International Conference on Coastal Engineering, Copenhagen, Denmark*. ASCE, New York, pp. 2750-2758.

Groeneweg, J. and Klopman, G., 1998. Changes of the mean velocity profiles in the combined wave-current motion in a GLM formulation, *J. Fluid Mechanics, Vol. 370*, pp. 271-296.

Groeneweg, J., 1999. *Wave-current interactions in a generalised Lagrangian Mean formulation*. PhD thesis, Delft University of Technology, Delft.

Grunnet, N.M., Walstra, D.J.R., and Ruessink, B.G., 2004. Process-based modelling of a shoreface nourishment. *Coastal Engineering, Vol. 51* pp. 581-607

Grunnet, N.M.; Ruessink, B.G.; Walstra, D.J.R., 2005. The influence of tides,wind and waves on the redistribution of nourished sediment, Terschelling, the Netherlands. *Coastal Engineering, Volume 52*, p.617-631

Hands, E.B., and Shepsis, V., 1999. Cyclic channel movement at the entrance to Willapa Bay, Washington, U.S.A. *Proceedings of Coastal Sediments '99*, pp. 1522-1536.

Hjelmfelt, A.T. and Lenau, C.W., 1970. Nonequilibrium transport of suspended sediment. *Journal Hydraulics Division*, ASCE, July 1970, pp. 1567-1586

Hoitink, A.J.F., Hoekstra, P., and van Maren, D.S., 2003. Flow asymmetry associated with astronomical tides: Implications for the residual transport of sediment, *J. Geophysical Research, 108(C10)*, 3315.

Holthuijsen, L.H., Booij, N., and Herbers, T.H.C., 1989. A prediction model for stationary, short-crested waves in shallow water with ambient currents. *Coastal Engineering, 13*, pp. 23-54.

Holthuijsen, L.H., Booij, N., and Ris, R.C., 1993. A spectral wave model for the coastal zone. *Proc. of the 2nd Int. Symposium on Ocean Wave Measurement and Analysis*, New Orleans, 630-641.

Huang, W. and Spaulding, M., 1996. Modelling horizontal diffusion with sigma coordinate system, *Journal of Hydraulic Engineering, Vol. 122*, No. 6, 349-352.

Ikeda, S., 1982. Lateral Bed-load Transport on Side Slopes. *Journal Hydraulics Division, ASCE, Vol. 108*, No. 11.

Isobe, M. and Horikawa, K., 1982. Study on water particle velocities of shoaling and breaking waves. *Coastal Engineering in Japan, Vol. 25*.

Johnson, H.K., Broker, I., and Zyserman, J.A., 1994. Identification of some relevant processes in coastal morphological modelling. *Coastal Engineering 1994*, pp. 2871-2885.

Kaminsky, G.M., Daniels, R.C., Huxford, R., McCandless, D., and Ruggiero, P., 1999. Mapping erosion hazard areas in Pacific County, Washington. *Journal of Coastal Research Special Issue 28*, p. 158-170.

Kolmogorov, A. N., 1942. Equations of turbulent motion of an incompressible fluid. IZV Akad. Nauk. USSR, Ser. Phys., Vol. 6, pp. 56-58. (translated into English by D.B. Spalding, as *Imperial College, Mechanical Engineering Department Report ON/6*, 1968, London, U.K.).

Kraus, N.C., Kurrus, K., Militello, A., Phillips, S., Scheffner, N.W., Seabergh, W.C., Shepsis, V., Smith, J.M., and Titus, C., 2000. *Study of Navigation Channel Feasibility, Willapa Bay, Washington*. Kraus, N. C. (Ed.). US Army Research and Development Center Technical Report ERDC/CHL TR-00-6.

Kreeke, van de J., and Robaczewska, K., 1993. Tide-induced residual transport of coarse sediment; application to the Ems estuary. *Netherlands Journal of Sea Research 31* (3), pp. 209-220.

Latteux, B., 1995. Techniques for long-term morphological simulation under tidal action. *Marine Geology, 126*, pp 129-141.

Leendertse, J.J., 1987. *A three-dimensional alternating direction implicit model with iterative fourth order dissipative non-linear advection terms.* WD-333-NETH, The Netherlands Rijkswaterstaat.

Ledden, van M., 2001. *Modelling of sand-mud mixtures. Part II: A process-based sand-mud model.* WL|Delft Hydraulics Report Z2840, The Netherlands. Delft Hydraulics, Delft, The Netherlands.

Ledden, van M. and Wang, Z.B., 2001. Sand-mud morphodynamics in an estuary. *River, Coastal and Estuarine Morphodynamics Conference RCEM2001 (IAHR), Japan.* Springer, New York, pp. 505-514.

Lesser, G.R., 2000. *Computation of Three-dimensional Suspended Sediment Transport within the DELFT3D-FLOW Module.* MSc thesis No. HE066, IHE, Delft, The Netherlands

Lesser, G.R., de Vroeg, J.H., Roelvink, J.A., de Gerloni, M., Ardone, V., 2003. Modelling the morphological impact of submerged offshore Breakwaters. *Proc. Coastal Sediments V 03.*

Lesser, G.R., Roelvink, J.A., van Kester, J.A.T.M., Stelling, G.S., 2004. Development and validation of a three-dimensional morphological model. *Coastal Engineering 51* (8-9), 883-915 (October).

Luepke, G., and Clifton, H.E., 1983. Heavy-mineral distribution in modern and ancient bay deposits, Willapa Bay, Washington. *Sedimentary Geology, v. 35*, p. 233-247.

Mellor, G.L. and Blumberg, A.F., 1985. Modelling vertical and horizontal diffusivities and the sigma coordinate system. *Monthly Weather Review, Vol. 113*, 1379-1383.

Morton, R.A., Purcell, N.A., and Peterson, R.L., 2002. *Large-scale cylcles of Holocene deposition and erosion at the entrance to Willapa Bay, Washington: implications for future land loss and coastal change.* U.S. Geological Survey Open-file Report 02-46, 124 pp.

Nicholson, J., Broker, I., Roelvink, J.A., Price, D., Tanguy, J.M., Moreno, L., 1997. Intercomparison of coastal area morphodynamics models. *Coastal Engineering, Vol. 31*, p. 97-123

Olsen, N.R.B., 2003. Three-Dimensional CFD Modeling of Self-Forming Meandering Channel. *J. Hydraulic Engineering, Vol. 129*, 366.

Overeem, van J., Steijn, R.C. and van Banning, G.K.F.M., 1992. Simulation of morphodynamics of tidal inlet in the Wadden Sea. In: Sterr, H., Hofstede, J., and Plag, H., (Editors), *Proceedings International Coastal Congress, Kiel*, Peter Lang Verlag, Frankfurt am Main, pp. 351-364.

Partheniades, E., 1965. *Erosion and Deposition of Cohesive Soils*. ASCE Journal of Hydraulic Division 91 (HY1), 105-139.

Péchon, P. and Teisson, C., 1996. Numerical modelling of bed evolution behind a detached breakwater. *Proceedings 25th International Conference on Coastal Engineering*, ASCE, New York, pp. 2050-2057.

Prandtl, L., 1945. *Uber ein neues formelsystem fur die ausgebildete turbulenz (On a new formation for fully developed turbulence)*. Nachrichten der Akademie der Wissenschaften (Report of Academy of Sciences, Gottingen, Germany), 6-19.

Pugh, D., 1987. *Tides, Surges and Mean Sea-Level*, 472 pp., John Wiley, Hoboken, N.J.

Ray, R., 1999. *A global ocean tide model from Topex/Poseidon altimetry: GOT99.2.* NASA Tech Memo 209478, 58 pages, Sept. 1999.

Reniers, A.J.H.M., Roelvink, J.A., Thornton, E.B., 2004. Morphodynamic modelling of an embayed beach under wave group forcing. *Journal of Geophysical Research 109* (C01030).

Rijn, van L.C., 1984. Sediment transport, Part II: Suspended Load Transport. *Journal of Hydraulic Engineering, No. 12.*

Rijn, van L.C., 1987. *Mathematical Modelling of Morphological Processes in the case of Suspended Sediment Transport.* Delft Technical University, Delft, The Netherlands. (also Delft Hydraulics Communication, No 382).

Rijn, van L.C., 1993. *Principles of sediment transport in Rivers, Estuaries and Coastal Seas.* Aqua Publications, Amsterdam.

Rijn, van L.C., 2001. *Approximation formulae for sand transport by currents and waves and implementation in DELFT-MOR.* WL|Delft Hydraulics Report Vol. Z3054.20. Delft Hydraulics, The Netherlands (unpublished report for Rijkswaterstaat/RIKZ).

Rijn, van L.C., Walstra DJR, Grasmeijer B, Sutherland J, Pan S, Sierra JP, 2003. The predictability of cross-shore bed evolution of sandy beaches at the time scale of storms and seasons using process-based profile models. *Coastal Engineering 47*, pp 295-327

Ris, R.C., Booij, N., and Holthuisen, L.H., 1999. A third-generation wave model for coastal regions, Part II: Verification. *J. Geophysical Research, 104*, C4, 7667-7681.

Rodi, W., 1984. *Turbulence models and their application in Hydraulics, State-of-the-art paper article sur l'etat de connaissance.* Paper presented by the IAHR-Section on Fundamentals of Division II: Experimental and Mathematical Fluid Dynamics, The Netherlands.

Roelvink, J.A., van Banning, G.K.F.M., 1994. Design and Development of Delft-3D and application to coastal morphodynamics. In: Verwey, A., Minns, A.W., Babovic, V., Maksimovic, M. Eds., *Hydroinformatics '94*. Balkema, Rotterdam, pp. 451-456.

Roelvink, J.A., Walstra, D.J.R., Chen, Z., 1994. Morphological modelling of Keta lagoon case. *Proceedings of the 24th International Conference on Coastal Engineering*. ASCE, Kobe, Japan.

Roelvink, J.A., Alain B., and Stam, J.M.T., 1998. A simple method to predict long-term morphological changes. *Proceedings of the 26th International Conference on Coastal Engineering, Copenhagen*, ASCE, New York, pp. 3224-3237.

Roelvink, J.A., Jeuken, M.C.J.L., van Holland, G., Aarninkhof, S.G.L. and Stam, J.M.T., 2001. Long-term, process-based modelling of complex areas. *Proceedings of the 4th Coastal Dynamics Conference, Lund, Sweden*. ASCE, New York, pp. 383-392.

Roelvink, J.A., Kessel, van T., Alfageme, S., and Canizares, R., 2003. Modelling of barrier island response to storms. Proc. *Proc. Coastal Sediments V 03.*

Roelvink J.A., 2006. Coastal morphodynamic evolution techniques. *Coastal Engineering 53*, pp 277-287

Roelvink J.A., Lesser, G.R., and Wegen, van der M., 2006. Morphological modelling of the wet-dry interface at various timescales. *Proceedings of the 7th International Conference on HydroScience and Engineering Philadelphia, USA September 10-13, 2006 (ICHE 2006)*. Drexel University, College of Engineering. ISBN: 0977447405.

Ruessink, G., Miles, J., Feddersen, F., Guza, R.T., Elgar, S., 2001. Modeling the alongshore current on barred beaches. *J. Geophysical Research 106*, 22451-22463.

Ruggiero, P., Kaminsky, G.M., Gelfenbaum, G., Voigt, B., 2005. Seasonal to Inter-annual Morphodynamics along a High-Energy Dissipative Littoral Cell. *Journal of Coastal Research, Vol. 21*, No. 3 pp. 553-578.

Simonin, O., Uittenbogaard, R.E., Baron, F., and Viollet, P.L., 1989. Possibilities and limitations to simulate turbulence fluxes of mass and momentum, measured in a steady stratified mixing layer. *Proceedings XXIII IAHR Congress, Ottawa, August 21-25*. National Research Council Canada, pp. A55-A62.

Soulsby, R.L., Hamm, L., Klopman, G., Myrhaug, D., Simons, R.R., and Thomas, G.P., 1993. Wave-current interaction within and outside the bottom boundary layer. *Coastal Engineering, 21* pp 41-69, Elsevier Science Publishers B.V., Amsterdam.

SPM, 1984. *Shore Protection Manual.* Waterways Experiment Station, US Army Corps of Engineers, Vicksburg, MS.

Steijn, R.C., 1992. *Input filtering techniques for complex morphological models.* Delft Hydraulics, Rept. H 824.53.

Steijn, R., Roelvink, J.A., Rakhorst, D., Ribberink, J., and van Overeem, J., 1998. North Coast of Texel: a comparison between reality and prediction. *Proc. 26th Int. Conf. On Coastal Engineering, Copenhagen,* ASCE, New York, pp. 2281-2293.

Stelling, G.S. and Leendertse, J.J., 1991. Approximation of Convective Processes by Cyclic ADI methods. *Proceeding of the 2nd ASCE Conference on Estuarine and Coastal Modelling, Tampa.* ASCE, New York, pp. 771-782.

Stelling, G.S. and van Kester, J.A.T.M., 1994. On the approximation of horizontal gradients in sigma coordinates for bathymetry with steep bottom slopes. *International Journal of Numerical Methods in Fluids, Vol. 18*, 915-955.

Stive, M.J.F. and Wind, H.G., 1986. Cross-shore mean flow in the surf zone. *Coastal Engineering, Vol. 10*, pp. 235-340.

Stive, M.J.F. and Vriend, de H.J., 1995. Modelling shoreface profile evolution. *Marine Geology, 126*, pp. 235-248.

Stive, M.J.F. and Wang, Z.B., 2003. Morphodynamic modelling of tidal basins and coastal inlets. *Advances in coastal modelling, C. Lakhan (Ed.)* Elsvier Sciences, pp. 367-392.

Struiksma, N., 1983. *Results of movable bed experiments in the DHL curved flume, Report on Experimental Investigation.* TOW Report R657 - XVIII/M1771, Delft Hyd. Lab., The Netherlands.

Struiksma, N., Olesen, K.W., Flokstra, C. and de Vriend, H.J., 1984. Bed deformation in Curved Alluvial Channels. *Journal of Hydraulic Research, Vol. 23*, No. 1.

Struiksma, N., 1985. Prediction of 2-D Bed Topography in Rivers. *Journal of Hydraulic Engineering, ASCE, Vol. 111*, No. 8.

Sutherland, J., Peet, A.H. and Soulsby, R.L., 2004. Evaluating the performance of morphological models. *Coastal Engineering 51*, pp. 917-939.

Svendsen, I.A., 1985. On the formulation of the cross-shore wave-current problem. *Proceedings Workshop 'European Coastal Zones'.* National Technical University of Athens, Athens pp. 1.1-1.9.

Terich, T., and Levenseller, T., 1986. The severe erosion of Cape Shoalwater, Washington: *Journal of Coastal Research, v. 2*, p. 465-477.

Tillotson, K., and Komar, P., 1997. The wave climate of the Pacific Northwest (Oregon and Washington): a compari- son of data sources, *Journal of Coastal Research, 13*, no. 2, pp. 440-452.

Uittenbogaard, R.E., 1998. *Model for eddy diffusivity and viscosity related to subgrid velocity and bed topography.* (unpublished note).

Vossen, van B., 2000. *Horizontal Large Eddy Simulations; evaluation of computations with DELFT3D-FLOW.* Report MEAH-197, Delft University of Technology.

Verboom, G.K., and Slob, A., 1984. Weakly reflective boundary conditions for two-dimensional water flow problems. 5th International Conference on Finite Elements in Water Resources, June 1984, Vermont, *Advances in Water Resources, Vol. 7.* Delft Hydraulics, Delft, The Netherlands. Delft Hydraulics Publication No. 322.

Vriend, de H.J., 1987. 2DH Mathematical Modelling of Morphological Evolutions in Shallow Water. *Coastal Engineering 11*, 1-27

Vriend, de H.J., Zyserman, J., Nicholson, J, Roelvink, J.A., Pechon, P., and Southgate, H.N., 1993. Medium-term 2DH coastal area modelling. *Coastal Engineering, 21* pp. 193-224.

Walstra, D.J.R., Roelvink, J.A., and Groeneweg, J., 2000. Calculation of wave-driven currents in a 3D mean flow model. *Coastal Engineering 2000, Billy Edge (ed.), Vol. 2*, ASCE, New York, pp. 1050-1063.

Wang, Z.B., Louters, T., and Vriend, de H.J., 1995. Morphodynamic modelling for a tidal inlet in the Wadden Sea. *Marine Geology, 126*, pp. 289-300.

Watanabe, A. et al., 1986. Numerical prediction model of three-dimensional beach deformation around a structure. *Coastal Engineering In Japan, Vol. 29*, pp. 179-194.

Wegen, van der M., Thanh, D.Q., and Roelvink, J.A., 2006. Bank erosion and morphodynamic evolution in alluvial estuaries using a process based 2D model. *Proceedings of the 7th International Conference on HydroScience and Engineering Philadelphia, USA September 10-13, 2006 (ICHE 2006).* Drexel University, College of Engineering. ISBN: 0977447405.

Wegen, van der M., and Roelvink, J.A., 2008. Long-term morphodynamic evolution of a tidal embayment using a two-dimensional, process-based model, *J. Geophysical Research, 113*, C03016

Westhuysen, A. and Lesser, G.R., 2007. *Evaluation and development of wave-current interaction in SWAN.* Activity 6.4 of SBW project Waddenzee. Delft Hydraulics Report H4918.60 for Rijkswaterstaat/RIKZ. November 2007.

White, S.M., 1970. Mineralogy and geochemistry of continental shelf sediments off the Washington-Oregon coast: *Journal Sedimentary Petrology, v. 40*, p. 38-54.

Winter, C., Chiou, M.D., Riethmüller, R., Ernstsen, V.B., Hebbeln, D., and Flemming, B.W., 2006. The concept of "representative tides" in morphodynamic numerical modelling. *Geo-Marine Letters, Springer, Volume 26* pp. 125-132

Zyserman, J.A and Johnson, H.K., 2002. Modelling morphological processes in the vicinity of shore-parallel breakwaters. *Coastal Engineering, 45* pp. 261-284.

List of Figures

List of Tables

Appendices

Appendix A

Seasonal Wave Climate Schematisation

This appendix contains the schematisation of the wave and wind climates used for the 5-year morphological simulations described in Chapter 4. The revised schematisation used in Chapter 5 used identical wave classes, but representative winds were revised on the basis of equivalent wind stress. Morfac values also changed in the latter schematisation due to the doubling of the length of the morphological tide.

Table A.1 – *Winter 1998/99 (duration 181 days).*

Wave Class (°)	Rep. Dirn. (°)	Rep. H_s (m)	Rep. Period (s)	Occ. (%)	Wind Spd. (m/s)	Wind Dirn. (°)	Morfac (-)
$H_s < 1.2m$							
180 - 360	278	1.00	10.6	5.9	4.6	145	10.31
$1.2m \leq H_s < 3.0m$							
180 - 240	217	2.43	8.2	3.5	9.8	156	6.12
240 - 270	259	2.20	10.6	9.0	5.1	171	15.71
270 - 285	278	2.13	12.1	16.0	4.6	151	28.00
285 - 360	293	2.13	11.7	11.2	3.6	135	19.62
$3.0m \leq H_s < 5.0m$							
180 - 240	223	4.00	10.0	10.1	10.7	177	17.65
240 - 270	258	4.02	12.0	11.3	9.2	204	19.82
270 - 285	278	3.98	13.6	12.4	7.1	204	21.63
285 - 360	289	3.88	14.9	9.1	6.8	209	15.99
$5.0m \leq H_s < 9.0m$							
180 - 230	218	5.88	11.4	3.1	15.4	189	5.36
230 - 270	248	5.92	13.4	4.4	11.5	207	7.65
270 - 360	282	5.69	15.8	4.0	6.0	221	7.05

Table A.2 – *Summer 1999 (duration 184 days)*

Wave Class (°)	Rep. Dirn. (°)	Rep. H_s (m)	Rep. Period (s)	Occ. (%)	Wind Spd. (m/s)	Wind Dirn. (°)	Morfac (-)
$H_s < 0.8m$							
180 - 360	275	0.71	9.5	4.9	2.3	336	8.79
$0.8m \leq H_s < 2.5m$							
180 - 240	215	1.76	8.1	3.4	5.8	187	6.01
240 - 270	261	1.57	9.6	8.6	2.7	273	15.24
270 - 290	282	1.57	9.5	36.7	4.3	321	65.35
290 - 360	299	1.50	9.1	35.5	4.7	335	63.25
$2.5m \leq H_s < 9.0m$							
180 - 270	246	3.89	11.8	3.2	5.6	198	5.64
270 - 360	288	3.08	11.4	7.6	3.4	299	13.55

Table A.3 – *Winter 1999/2000 (duration 181 days).*

Wave Class (°)	Rep. Dirn. (°)	Rep. H_s (m)	Rep. Period (s)	Occ. (%)	Wind Spd. (m/s)	Wind Dirn. (°)	Morfac (-)
$H_s < 1.2m$							
180 - 360	269	1.02	11.0	7.6	6.5	218	13.33
$1.2m \leq H_s < 3.0m$							
180 - 230	218	2.22	8.8	4.5	7.7	220	7.93
230 - 250	239	2.23	9.5	5.8	5.3	213	10.19
250 - 270	261	2.10	11.1	8.9	4.1	213	15.58
270 - 285	279	2.20	14.0	27.0	2.6	206	47.56
285 - 360	290	2.12	12.8	10.3	3.6	205	18.12
$3.0m \leq H_s < 5.0m$							
180 - 240	223	3.85	9.5	5.7	9.8	195	9.99
240 - 270	259	3.84	11.5	8.5	6.8	218	14.98
270 - 285	278	3.83	14.1	13.6	5.2	217	23.92
285 - 360	290	3.75	14.0	5.2	6.8	197	9.14
$5.0m \leq H_s < 9.0m$							
180 - 270	245	5.57	12.4	1.3	10.0	208	2.26
270 - 360	283	5.47	14.7	1.6	7.6	226	2.90

Table A.4 – *Summer 2000 (duration 184 days).*

Wave Class (°)	Rep. Dirn. (°)	Rep. H_s (m)	Rep. Period (s)	Occ. (%)	Wind Spd. (m/s)	Wind Dirn. (°)	Morfac (-)
$H_s < 0.8m$							
180 - 360	275	0.66	10.0	7.5	3.6	395	13.29
$0.8m \leq H_s < 2.5m$							
180 - 240	213	1.62	7.9	6.0	5.6	198	10.71
240 - 270	259	1.60	9.6	10.1	2.4	232	18.06
270 - 290	281	1.55	9.9	23.0	3.2	329	40.97
290 - 360	300	1.56	8.9	42.2	4.8	334	75.36
$2.5m \leq H_s < 9.0m$							
180 - 270	238	3.55	10.8	4.6	7.5	173	8.24
270 - 360	287	3.18	12.6	6.3	2.6	210	11.19

Table A.5 – *Winter 2000/2001 (duration 181 days).*

Wave Class (°)	Rep. Dirn. (°)	Rep. H_s (m)	Rep. Period (s)	Occ. (%)	Wind Spd. (m/s)	Wind Dirn. (°)	Morfac (-)
$H_s < 1.2m$							
180 - 360	273	1.02	10.5	6.7	5.5	155	11.76
$1.2m \leq H_s < 3.0m$							
180 - 230	214	2.25	8.4	5.8	8.7	149	10.19
230 - 250	239	2.26	9.1	2.9	6.5	157	5.07
250 - 270	262	2.28	11.5	6.4	6.3	158	11.28
270 - 285	279	2.24	13.4	26.4	4.8	152	46.12
285 - 360	292	2.14	12.0	25.6	3.4	159	44.87
$3.0m \leq H_s < 5.0m$							
180 - 240	221	3.74	9.1	2.9	10.3	203	5.07
240 - 270	258	3.82	12.5	5.0	6.8	202	8.70
270 - 285	279	3.79	14.7	9.9	4.2	170	17.24
285 - 360	290	3.61	13.3	6.0	4.4	190	10.55
$5.0m \leq H_s < 9.0m$							
180 - 270	257	5.67	13.1	0.8	9.6	209	1.33
270 - 360	279	5.67	14.8	1.6	8.7	211	2.74

Table A.6 – *Summer 2001 (duration 184 days).*

Wave Class (°)	Rep. Dirn. (°)	Rep. H_s (m)	Rep. Period (s)	Occ. (%)	Wind Spd. (m/s)	Wind Dirn. (°)	Morfac (-)
$H_s < 0.8m$							
180 - 360	275	0.70	7.7	5.1	2.6	321	9.03
$0.8m \leq H_s < 2.5m$							
180 - 240	217	1.65	7.5	2.8	6.2	197	5.04
240 - 270	261	1.41	8.6	9.9	3.2	290	17.69
270 - 290	281	1.64	10.0	32.3	3.2	315	57.42
290 - 360	300	1.57	9.0	36.4	5.3	331	64.80
$2.5m \leq H_s < 9.0m$							
180 - 270	225	2.98	8.2	2.7	9.0	209	4.88
270 - 360	288	3.64	13.7	10.7	4.0	264	18.98

Table A.7 – *Winter 2001/2002 (duration 181 days).*

Wave Class (°)	Rep. Dirn. (°)	Rep. H_s (m)	Rep. Period (s)	Occ. (%)	Wind Spd. (m/s)	Wind Dirn. (°)	Morfac (-)
$H_s < 1.2m$							
180 - 360	275	1.06	9.3	7.2	2.5	380	12.56
$1.2m \leq H_s < 3.0m$							
180 - 230	212	2.06	8.1	4.7	8.0	186	8.32
230 - 250	242	2.24	9.7	4.1	6.8	141	7.19
250 - 270	261	2.28	11.5	6.8	4.7	127	11.91
270 - 285	278	2.24	12.5	15.1	3.6	131	26.50
285 - 360	295	2.13	10.9	22.9	3.9	321	40.11
$3.0m \leq H_s < 5.0m$							
180 - 240	223	3.94	10.4	7.4	11.1	201	13.05
240 - 270	257	3.84	12.5	8.9	8.2	208	15.63
270 - 285	277	3.80	13.2	8.3	5.2	215	14.58
285 - 360	294	3.81	13.0	8.9	6.0	244	15.67
$5.0m \leq H_s < 9.0m$							
180 - 270	242	5.81	13.2	3.1	12.6	202	5.41
270 - 360	283	6.11	14.6	2.3	7.4	252	4.00

Table A.8 – *Summer 2002 (duration 184 days).*

Wave Class (°)	Rep. Dirn. (°)	Rep. H_s (m)	Rep. Period (s)	Occ. (%)	Wind Spd. (m/s)	Wind Dirn. (°)	Morfac (-)
$H_s < 0.8m$							
180 - 360	278	0.71	9.2	9.0	3.9	336	16.03
$0.8m \leq H_s < 2.5m$							
180 - 240	217	1.48	8.0	4.7	6.3	202	8.34
240 - 270	259	1.35	9.4	10.3	3.8	326	18.25
270 - 290	281	1.60	11.1	27.6	3.5	346	49.14
290 - 360	300	1.56	8.8	43.5	4.9	331	77.45
$2.5m \leq H_s < 9.0m$							
180 - 270	214	2.96	8.9	0.5	9.9	172	0.89
270 - 360	292	2.78	11.5	4.3	5.1	328	7.73

Table A.9 – *Winter 2002/2003 (duration 181 days).*

Wave Class (°)	Rep. Dirn. (°)	Rep. H_s (m)	Rep. Period (s)	Occ. (%)	Wind Spd. (m/s)	Wind Dirn. (°)	Morfac (-)
$H_s < 1.2m$							
180 - 360	272	0.96	11.5	6.5	4.0	103	11.44
$1.2m \leq H_s < 3.0m$							
180 - 230	218	2.40	8.5	4.1	6.7	167	7.21
230 - 250	240	2.21	9.7	5.8	5.4	196	10.11
250 - 270	263	2.07	11.8	13.7	4.0	131	24.04
270 - 285	278	2.19	13.6	23.3	2.3	152	40.67
285 - 360	291	2.17	11.9	13.9	2.7	289	24.24
$3.0m \leq H_s < 5.0m$							
180 - 240	225	3.91	10.4	6.5	9.4	201	11.44
240 - 270	258	3.86	12.7	9.9	6.4	192	17.40
270 - 285	277	3.73	13.8	9.2	5.9	184	16.11
285 - 360	290	3.36	12.5	2.3	4.8	376	3.95
$5.0m \leq H_s < 9.0m$							
180 - 270	244	5.75	13.7	3.9	12.6	208	6.85
270 - 360	274	5.64	15.6	0.9	10.7	171	1.49

Table A.10 – *Summer 2003 (duration 184 days).*

Wave Class (°)	Rep. Dirn. (°)	Rep. H_s (m)	Rep. Period (s)	Occ. (%)	Wind Spd. (m/s)	Wind Dirn. (°)	Morfac (-)
$H_s < 0.8m$							
180 - 360	265	0.70	9.4	5.5	2.3	327	9.87
$0.8m \leq H_s < 2.5m$							
180 - 240	222	1.40	8.0	4.1	5.1	207	7.21
240 - 270	260	1.53	9.6	15.5	2.3	301	27.64
270 - 290	281	1.59	10.0	25.0	4.1	333	44.53
290 - 360	301	1.51	8.5	39.6	5.4	340	70.52
$2.5m \leq H_s < 9.0m$							
180 - 270	240	3.90	10.3	3.8	10.8	197	6.85
270 - 360	283	4.16	14.0	6.3	5.2	233	11.20

About the Author

Giles Lesser was born in Wellington, New Zealand on the 28th of May 1970. He studied civil engineering at the University of Canterbury in Christchurch, New Zealand from 1988 until 1991 graduating with first class honours. From 1992 until 1998 he worked as a civil engineer for Wellington City Council, initially in the Roading Asset Management section and later in the Roading Design section where he was involved with the design, construction, and maintenance of roads, sewer and stormwater drainage and urban landfill development. For a period he was also responsible for the maintenance of the city's many kilometres of coastal defences.

After leaving New Zealand in 1998, Giles travelled to Europe where he studied coastal engineering at IHE Delft (now UNESCO-IHE) where, in 2000, he was awarded an MSc with distinction. His thesis project, undertaken at Delft Hydraulics, involved the implementation of suspended sediment transport in the Delft3D computer package. Giles was employed by Delft Hydraulics as a researcher/advisor from 2001 until 2006. Initially based in Delft and living in Rotterdam he worked on a number of large-scale coastal development projects including morphological and water quality studies for the design of Palm Island Jumeirah (UAE) and ship motion and manoeuvring studies for Maasvlakte 2 (the second major extension of the Port of Rotterdam) and the Port of Dunkirk master plan.

In September 2002 Giles was seconded to the US Geological Survey and moved to Menlo Park, CA, USA. There he assumed the role of Visiting Scientist, assisting the USGS conduct several morphological modelling projects and field measurement campaigns including the entrance to Willapa Bay, discussed in this thesis, and the mouth of the Columbia River. He also worked with local consultants on modelling projects related to the restoration of salt ponds in San Francisco Bay.

In February 2006 Giles relocated to Melbourne, Australia where he now lives with his wife Lauren, two daughters Katherine and Evelyn, and new son Arthur. There he completed the work for the final chapter of this thesis while working part time for his present employer OMC International where he manages the Research Section, improving methods of predicting the under-keel clearance of vessels in draft-restricted channels.